180 Days of GEOGRAPHY

for Sixth Grade

Author

Jennifer Edgerton, Ed.M.

SHELL EDUCATION

Series Consultant

Nicholas Baker, Ed.D.
Supervisor of Curriculum and Instruction
Colonial School District, DE

Publishing Credits

Corinne Burton, M.A.Ed., *Publisher*
Conni Medina, M.A.Ed., *Managing Editor*
Emily R. Smith, M.A.Ed., *Content Director*
Veronique Bos, *Creative Director*
Shaun N. Bernadou, *Art Director*
Lynette Ordoñez, *Editor*
Marc Pioch, M.A.Ed., *Editor*
Kevin Pham, *Graphic Designer*
Stephanie Bernard, *Associate Editor*

Image Credits

Standards

For information on how this resource meets national and other state standards, see pages 10–14. You may also review this information by visiting our website at www.teachercreatedmaterials.com/administrators/correlations/ and following the on-screen directions.

Shell Education
A division of Teacher Created Materials
5301 Oceanus Drive
Huntington Beach, CA 92649-1030
www.tcmpub.com/shell-education

ISBN 978-1-4258-3307-7
©2018 Shell Educational Publishing, Inc.

TABLE OF CONTENTS

INTRODUCTION

With today's geographic technology, the world seems smaller than ever. Satellites can accurately measure the distance between any two points on the planet and give detailed instructions about how to get there in real time. This may lead some people to wonder why we still study geography.

While technology is helpful, it isn't always accurate. We may need to find detours around construction, use a trail map, outsmart our technology, and even be the creators of the next navigational technology.

But geography is also the study of cultures and how people interact with the physical world. People change the environment, and the environment affects how people live. People divide the land for a variety of reasons. Yet no matter how it is divided or why, people are at the heart of these decisions. To be responsible and civically engaged, students must learn to think in geographical terms.

The Need for Practice

To be successful in geography, students must understand how the physical world affects humanity. They must not only master map skills but also learn how to look at the world through a geographical lens. Through repeated practice, students will learn how a variety of factors affect the world in which they live.

Understanding Assessment

In addition to providing opportunities for frequent practice, teachers must be able to assess students' geographical understandings. This allows teachers to adequately address students' misconceptions, build on their current understandings, and challenge them appropriately. Assessment is a long-term process that involves careful analysis of student responses from a discussion, project, practice sheet, or test. The data gathered from assessments should be used to inform instruction: slow down, speed up, or reteach. This type of assessment is called *formative assessment*.

HOW TO USE THIS BOOK

Weekly Structure

The first two weeks of the book focus on map skills. By introducing these skills early in the year, students will have a strong foundation on which to build throughout the year. Each of the remaining 34 weeks will follow a regular weekly structure.

Each week, students will study a grade-level geography topic and a location in the world. Locations may be a town, a state, a region, or the whole continent.

Days 1 and 2 of each week focus on map skills. Days 3 and 4 allow students to apply information and data to what they have learned. Day 5 helps students connect what they have learned to themselves.

 Day 1—Reading Maps: Students will study a grade-appropriate map and answer questions about it.

 Day 2—Creating Maps: Students will create maps or add to an existing map.

 Day 3—Read About It: Students will read a text related to the topic or location for the week and answer text-dependent or photo-dependent questions about it.

 Day 4—Think About It: Students will analyze a chart, diagram, or other graphic related to the topic or location for the week and answer questions about it.

 Day 5—Geography and Me: Students will do an activity to connect what they learned to themselves.

Five Themes of Geography

Good geography teaching encompasses all five themes of geography: location, place, human-environment interaction, movement, and region. Location refers to the absolute and relative locations of a specific point or place. The place theme refers to the physical and human characteristics of a place. Human-environment interaction describes how humans affect their surroundings and how the environment affects the people who live there. Movement describes how and why people, goods, and ideas move between different places. The region theme examines how places are grouped into different regions. Regions can be divided based on a variety of factors, including physical characteristics, cultures, weather, and political factors.

HOW TO USE THIS BOOK (cont.)

Weekly Themes

The following chart shows the topics, locations, and themes of geography that are covered during each week of instruction.

Wk.	Topic	Location	Geography Themes
1	—Map Skills Only—		Location
2			Location
3	Diversity	Africa	Location, Place, Region
4	The Nile River	Egypt (Africa)	Human-Environment Interaction
5	Changing Environment	Sahara Desert (Africa)	Place, Human-Environment Interaction
6	Transportation and Landforms	Congo River Basin (Africa)	Place, Human-Environment Interaction, Movement
7	Trade and Cultural Diffusion	East Africa	Movement, Region
8	Regions	Asia	Location, Region
9	Early Civilizations	River Valleys (Asia)	Place, Human-Environment Interaction, Region
10	Weather Patterns	South and Southeast Asia	Human-Environment Interaction
11	Population Growth and Urbanization	South and Southeast Asia	Place
12	Religion	Asia	Movement, Region
13	Migration	Oceania	Human-Environment Interaction, Movement
14	Continents	Zealandia (Australia/Oceania)	Region
15	Settlement	Australia	Place, Human-Environment Interaction
16	Vegetation and Wildlife	Australia	Location, Place
17	Coral Reefs	Great Barrier Reef (Australia/Oceania)	Place, Human-Environment Interaction
18	Cultural Diffusion	Greek and Roman Empires (Europe)	Movement, Region
19	Volcanoes	Vesuvius (Europe/Asia)	Human-Environment Interaction
20	Fjords	Scandinavia (Europe)	Place, Region

HOW TO USE THIS BOOK (cont.)

Wk.	Topic	Location	Geography Themes
21	Railroad Development	Europe	Movement
22	Countries and Subdivisions	British Isles (Europe)	Location, Region
23	Early Civilization	Mexico (North America)	Place, Human-Environment Interaction
24	Bodies of Water	Canada (North America)	Human-Environment Interaction
25	Wildlife Migration	North America	Human-Environment Interaction
26	Hurricanes	Caribbean (North America)	Human-Environment Interaction, Region
27	Exports	Central America (North America)	Human-Environment Interaction, Movement
28	Distance/Size	South America	Location, Place
29	Conservation	Amazon (South America)	Human-Environment Interaction
30	Early Civilization	Andes Mountains (South America)	Place, Human-Environment Interaction
31	Exports	Chile (South America)	Human-Environment Interaction, Movement
32	Conservation	Galápagos Islands (South America)	Place, Human-Environment Interaction
33	Fossil Fuels	World	Human-Environment Interaction, Movement
34	Renewable Energy	World	Human-Environment Interaction
35	Immigrants and Refugees	World	Movement
36	Sustainable Urban Development	World	Location, Human-Environment Interaction

HOW TO USE THIS BOOK *(cont.)*

Using the Practice Pages

The activity pages provide practice and assessment opportunities for each day of the school year. Teachers may wish to prepare packets of weekly practice pages for the classroom or for homework.

As outlined on page 4, each week examines one location and one geography topic.

The first two days focus on map skills. On Day 1, students will study a map and answer questions about it. On Day 2, they will add to or create a map.

Days 3 and 4 allow students to apply information and data from texts, charts, graphs, and other sources to the location being studied.

On Day 5, students will apply what they learned to themselves.

Using the Resources

Rubrics for the types of days (map skills, applying information and data, and making connections) can be found on pages 210–212 and in the Digital Resources. Use the rubrics to assess students' work. Be sure to share these rubrics with students often so that they know what is expected of them.

HOW TO USE THIS BOOK *(cont.)*

Diagnostic Assessment

Teachers can use the practice pages as diagnostic assessments. The data analysis tools included with the book enable teachers or parents to quickly score students' work and monitor their progress. Teachers and parents can quickly see which skills students may need to target further to develop proficiency.

Students will learn map skills, how to apply text and data to what they have learned, and how to relate what they learned to themselves. Teachers can assess students' learning in each area using the rubrics on pages 210–212. Then, record their scores on the Practice Page Item Analysis sheets on pages 213–215. These charts are also provided in the Digital Resources as PDFs, Microsoft Word® files, and Microsoft Excel® files (see page 216 for more information). Teachers can input data into the electronic files directly on the computer, or they can print the pages.

To Complete the Practice Page Item Analyses:

- Write or type students' names in the far-left column. Depending on the number of students, more than one copy of the forms may be needed.
  - The skills are indicated across the tops of the pages.
  - The weeks in which students should be assessed are indicated in the first rows of the charts. Students should be assessed at the ends of those weeks.

- Review students' work for the days indicated in the chart. For example, if using the Making Connections Analysis sheet for the first time, review students' work from Day 5 for all five weeks.

- Add the scores for each student. Place that sum in the far right column. Record the class average in the last row. Use these scores as benchmarks to determine how students are performing.

Digital Resources

The Digital Resources contain digital copies of the rubrics, analysis pages, and standards charts. See page 216 for more information.

HOW TO USE THIS BOOK *(cont.)*

Using the Results to Differentiate Instruction

Once results are gathered and analyzed, teachers can use them to inform the way they differentiate instruction. The data can help determine which geography skills are the most difficult for students and which students need additional instructional support and continued practice.

Whole-Class Support

The results of the diagnostic analysis may show that the entire class is struggling with certain geography skills. If these concepts have been taught in the past, this indicates that further instruction or reteaching is necessary. If these concepts have not been taught in the past, this data is a great preassessment and may demonstrate that students do not have a working knowledge of the concepts. Thus, careful planning for the length of the unit(s) or lesson(s) must be considered, and additional front-loading may be required.

Small-Group or Individual Support

The results of the diagnostic analysis may show that an individual student or a small group of students is struggling with certain geography skills. If these concepts have been taught in the past, this indicates that further instruction or reteaching is necessary. Consider pulling these students aside to instruct them further on the concepts while others are working independently. Students may also benefit from extra practice using games or computer-based resources.

Teachers can also use the results to help identify proficient individual students or groups of students who are ready for enrichment or above-grade-level instruction. These students may benefit from independent learning contracts or more challenging activities.

STANDARDS CORRELATIONS

Shell Education is committed to producing educational materials that are research and standards based. In this effort, we have correlated all our products to the academic standards of all 50 states, the District of Columbia, the Department of Defense Dependents Schools, and all Canadian provinces.

How to Find Standards Correlations

To print a customized correlation report of this product for your state, visit our website at **www.teachercreatedmaterials.com/administrators/correlations** and follow the on-screen directions. If you require assistance in printing correlation reports, please contact our Customer Service Department at 1-877-777-3450.

Purpose and Intent of Standards

The Every Student Succeeds Act (ESSA) mandates that all states adopt challenging academic standards that help students meet the goal of college and career readiness. While many states already adopted academic standards prior to ESSA, the act continues to hold states accountable for detailed and comprehensive standards. Standards are designed to focus instruction and guide adoption of curricula. Standards are statements that describe the criteria necessary for students to meet specific academic goals. They define the knowledge, skills, and content students should acquire at each level. Standards are also used to develop standardized tests to evaluate students' academic progress. Teachers are required to demonstrate how their lessons meet state standards. State standards are used in the development of our products, so educators can be assured they meet the academic requirements of each state.

The activities in this book are aligned to the National Geography Standards and the McREL standards. The chart on pages 11–12 lists the National Geography Standards used throughout this book. The chart on pages 13–14 correlates the specific McREL and National Geography Standards to each week. The standards charts are also in the Digital Resources (standards.pdf).

C3 Framework

This book also correlates to the College, Career, and Civic Life (C3) Framework published by the National Council for the Social Studies. By completing the activities in this book, students will learn to answer and develop strong questions (Dimension 1), critically think like a geographer (Dimension 2), and effectively choose and use geography resources (Dimension 3). Many activities also encourage students to take informed action within their communities (Dimension 4).

STANDARDS CORRELATIONS (cont.)

180 Days of Geography is designed to give students daily practice in geography through engaging activities. Students will learn map skills, how to apply information and data to their understandings of various locations and cultures, and how to apply what they learned to themselves.

Easy to Use and Standards Based

There are 18 National Geography Standards, which fall under six essential elements. Specific expectations are given for fourth grade, eighth grade, and twelfth grade. For this book, eighth grade expectations were used with the understanding that full mastery is not expected until that grade level.

Essential Elements	National Geography Standards
The World in Spatial Terms	**Standard 1:** How to use maps and other geographic representations, geospatial technologies, and spatial thinking to understand and communicate information
	Standard 2: How to use mental maps to organize information about people, places, and environments in a spatial context
	Standard 3: How to analyze the spatial organization of people, places, and environments on Earth's surface
Places and Regions	**Standard 4:** The physical and human characteristics of places
	Standard 5: People create regions to interpret Earth's complexity
	Standard 6: How culture and experience influence people's perceptions of places and regions
Physical Systems	**Standard 7:** The physical processes that shape the patterns of Earth's surface
	Standard 8: The characteristics and spatial distribution of ecosystems and biomes on Earth's surface

STANDARDS CORRELATIONS *(cont.)*

Essential Elements	National Geography Standards
Human Systems	**Standard 9:** The characteristics, distribution, and migration of human populations on Earth's surface
	Standard 10: The characteristics, distribution, and complexity of Earth's cultural mosaics
	Standard 11: The patterns and networks of economic interdependence on Earth's surface
	Standard 12: The process, patterns, and functions of human settlement
	Standard 13: How the forces of cooperation and conflict among people influence the division and control of Earth's surface
Environment and Society	**Standard 14:** How human actions modify the physical environment
	Standard 15: How physical systems affect human systems
	Standard 16: The changes that occur in the meaning, use, distribution, and importance of resources
The Uses of Geography	**Standard 17:** How to apply geography to interpret the past
	Standard 18: How to apply geography to interpret the present and plan for the future

–2012 National Council for Geographic Education

STANDARDS CORRELATIONS *(cont.)*

Easy to Use and Standards Based *(cont.)*

This chart lists the specific National Geography Standards (NGS) and McREL standards that are covered each week.

Wk.	NGS	McREL Standards
1	Standard 1	Knows the advantages and disadvantages of maps, globes, and other geographic tools to illustrate a data set.
2	Standard 1	Uses thematic maps. Knows the advantages and disadvantages of maps, globes, and other geographic tools to illustrate a data set.
3	Standards 9 and 10	Understands the factors that affect the cohesiveness and integration of countries.
4	Standards 12 and 15	Understands the ways in which technology influences the human capacity to modify the physical environment. Knows how the physical environment affects life in different regions.
5	Standards 8 and 17	Knows the causes and effects of changes in a place over time.
6	Standard 15	Knows the ways in which human systems develop in response to conditions in the physical environment.
7	Standard 11	Understands how places are connected and how these connections demonstrate interdependence and accessibility.
8	Standard 5	Understands criteria that give a region identity.
9	Standards 12 and 17	Knows the ways people take aspects of the environment into account when deciding on locations for human activities.
10	Standard 15	Knows how the physical environment affects life in different regions.
11	Standard 9	Understands demographic concepts and how they are used to describe population characteristics of a country or region.
12	Standard 10	Knows the social, political, and economic divisions on Earth's surface at the local, state, national, and international levels.
13	Standards 13 and 15	Knows the ways in which human systems develop in response to conditions in the physical environment.
14	Standard 7	Knows the major processes that shape patterns in the physical environment.
15	Standards 6 and 15	Knows the ways in which human systems develop in response to conditions in the physical environment.
16	Standard 8	Understands the environmental consequences of people changing the physical environment.
17	Standard 14	Knows the causes and effects of changes in a place over time. Understands the environmental consequences of people changing the physical environment.
18	Standards 10 and 17	Understands the significance of patterns of cultural diffusion.

STANDARDS CORRELATIONS *(cont.)*

Wk.	NGS	McREL Standards
19	Standards 7, 15, and 17	Knows the effects of natural hazards on human systems in different regions of the United States and the world.
20	Standard 7	Knows the physical characteristics of places.
21	Standard 8	Understands historic and contemporary systems of transportation and communication in the development of economic activities.
22	Standards 4 and 5	Understands criteria that give a region identity.
23	Standards 10 and 17	Understands the environmental consequences of people changing the physical environment.
24	Standard 16	Knows the ways in which culture influences the perception of places and regions.
25	Standards 8 and 14	Understands the functions and dynamics of ecosystems.
26	Standards 15 and 18	Knows the effects of natural hazards on human systems in different regions of the United States and the world.
27	Standard 11	Understands issues related to the spatial distribution of economic activities.
28	Standard 13	Knows the social, political, and economic divisions on Earth's surface at the local, state, national, and international levels.
29	Standard 14	Understands the environmental consequences of people changing the physical environment.
30	Standards 12 and 15	Knows how the physical environment affects life in different regions.
31	Standard 11	Understands the role of technology in resource acquisition and use, and its impact on the environment.
32	Standards 14 and 18	Understands the environmental consequences of people changing the physical environment.
33	Standards 14, 16 and 18	Knows world patterns of resource distribution and utilization.
34	Standards 16 and 18	Understands how the development and widespread use of alternative energy sources might have an impact on societies.
35	Standard 9	Knows the ways in which human movement and migration influence the character of a place.
36	Standards 14 and 18	Understands patterns of land use in urban, suburban, and rural areas.

Name: Brooklyn   **Date:** _____

Directions: A data frame is the purpose of the map. An easy way to show the data frame is to title the map. Draw a map of the world using the data frame of the continents and oceans. Then, title the map. Use the Word Bank to label the continents and oceans.

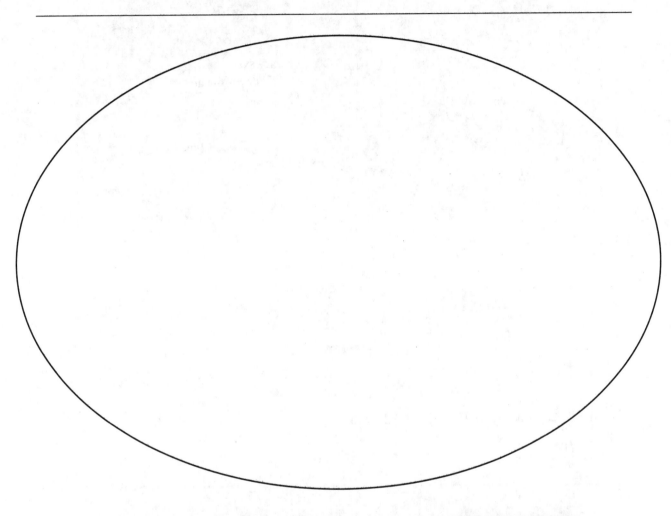

Word Bank		
Africa	Antarctica	Asia
Australia	Europe	North America
South America	Pacific Ocean	Atlantic Ocean
Indian Ocean	Arctic Ocean	Southern Ocean

Name: Brooklyn

Date: _____

Directions: Maps often use symbols to show information. Symbols can be figures, shapes, lines, and colors. A legend explains what the symbols mean. Use the map and its legend to answer the questions.

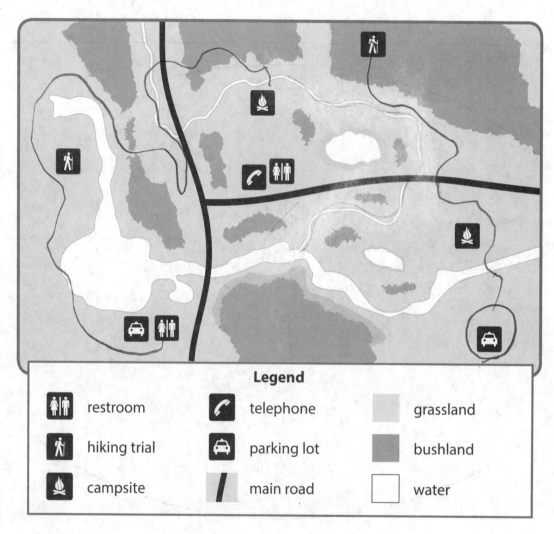

Legend

🚻	restroom	☎	telephone		grassland
🚶	hiking trial	🚗	parking lot		bushland
🔥	campsite	/	main road		water

1. What type of land feature makes up most of the park?

   Grassland

2. How many parking lots are shown on the map?

   2

3. Circle the parking lot that is farthest from the main roads.

4. Draw two different routes you could take to explore the area shown on the map.

Name: Brooklyn

Date: _____

Directions: Scales and compass roses help people understand maps. A scale shows how far a distance on a map is in real life. A compass rose shows cardinal directions (north, south, east, and west). Use the scale and compass rose to answer the questions.

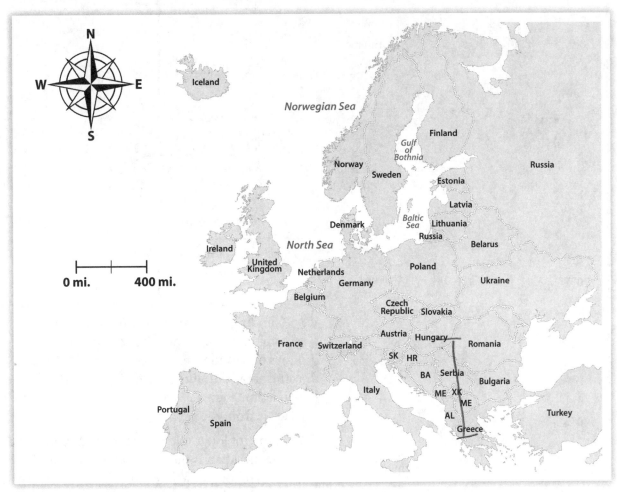

1. What is the approximate width of Poland from east to west?

   400mi

2. Assume that you can travel 400 miles with one tank of gas. If you had one tank of gas and could drive in a straight line, what route would allow you to visit the most countries? Draw it on the map.

3. Explain why the route you drew was the best one possible.

   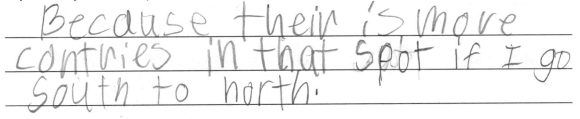
   Because their is more contries in that spot if I go south to north.

Name: Brooklyn   **Date:** _____

Directions: Read the text, and study the map. Then, answer the questions.

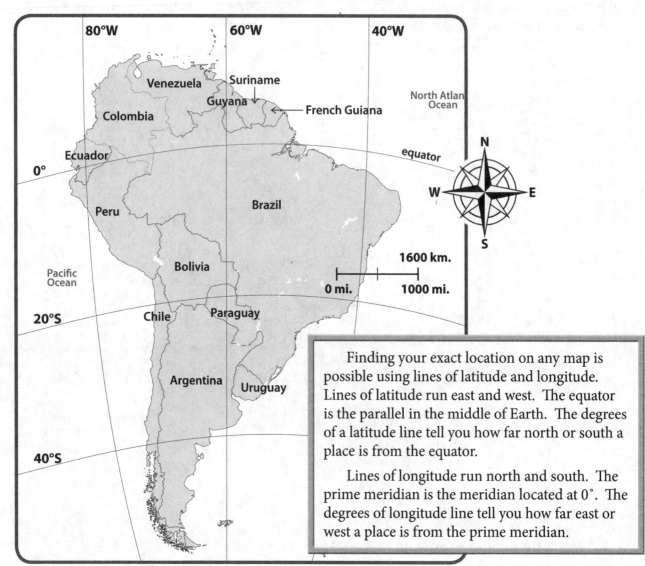

Finding your exact location on any map is possible using lines of latitude and longitude. Lines of latitude run east and west. The equator is the parallel in the middle of Earth. The degrees of a latitude line tell you how far north or south a place is from the equator.

Lines of longitude run north and south. The prime meridian is the meridian located at 0°. The degrees of longitude line tell you how far east or west a place is from the prime meridian.

1. What happens to the degrees of longitude as you move farther west on the map? Why?

_____

_____

2. What is the approximate distance between degrees of latitude on this map?

_____

3. What is the approximate distance between degrees of longitude on this map?

_____

Name: _____ **Date:** _____

Directions: Create a detailed map of your classroom. Include the following map elements: title, legend, compass rose, and symbols to indicate at least four items in your classroom.

Map Skills

Name: _____ **Date:** _____

Directions: A physical map shows the natural features of an area. Follow the steps to complete the physical map of India.

_____

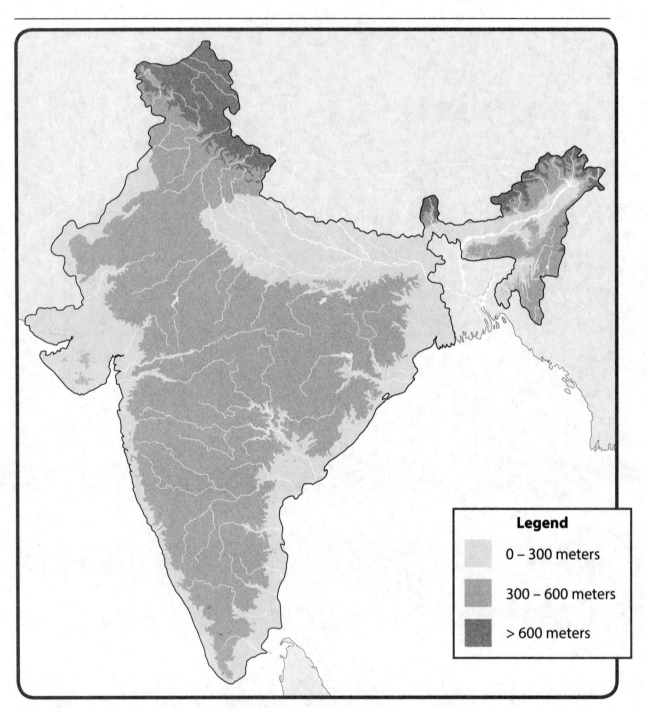

Legend

0 – 300 meters

300 – 600 meters

> 600 meters

1. Color all rivers, lakes, and oceans blue.

2. Circle the areas with the highest elevations on the map.

3. Title the map.

Name: _____ Date: _____

Directions: A political map shows borders between countries and states. Capital cities and other large cities may be labeled. Color the map using a different color for each country. Then, label the capital city in each country.

Country	Capital
Belize	Belmopan
Canada	Ottawa
Costa Rica	San José
El Salvador	San Salvador
Guatemala	Guatemala City

Country	Capital
Honduras	Tegucigalpa
Mexico	Mexico City
Nicaragua	Managua
Panama	Panama City
United States	Washington, DC

Map Skills

Name: _____ Date: _____

Directions: A thematic map highlights a theme or special topic. This thematic map shows the annual rainfall in different regions of China. Study the map, and answer the questions.

Rainfall in China

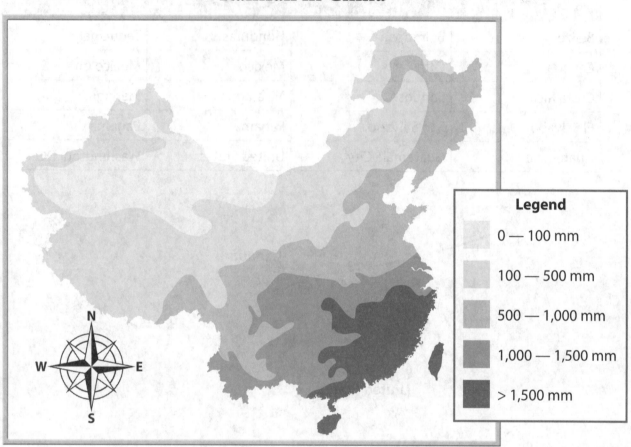

Legend

0 — 100 mm

100 — 500 mm

500 — 1,000 mm

1,000 — 1,500 mm

> 1,500 mm

1. How do you read this map?

_____

_____

2. Which areas in China have the highest annual rainfall?

_____

_____

3. Why might a farmer use this map?

_____

_____

Name: _____ Date: _____

Directions: Use the chart to create a thematic map of the Middle East showing the amount of oil from these countries. Add a legend to help readers interpret your map.

Oil in the Middle East

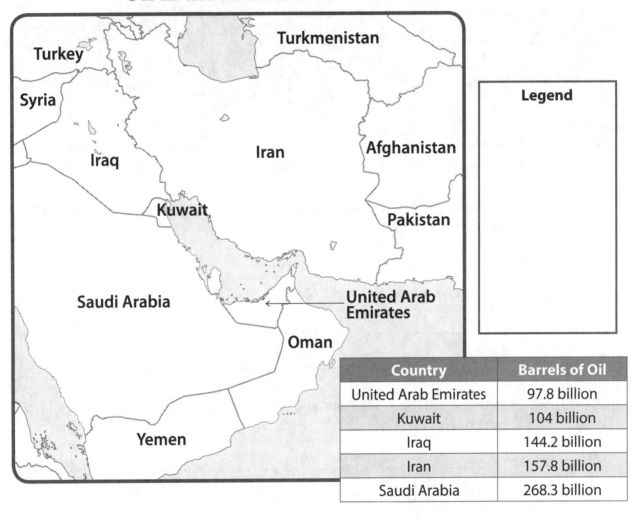

Country	Barrels of Oil
United Arab Emirates	97.8 billion
Kuwait	104 billion
Iraq	144.2 billion
Iran	157.8 billion
Saudi Arabia	268.3 billion

1. Explain why you chose to create your thematic map the way you did.

   _____

   _____

2. Does your map make it easier to understand the information from the chart? Why or why not?

   _____

   _____

   _____

Map Skills

Name: _____ Date: _____

Directions: Read each scenario, and decide whether you would use a political, physical, or thematic map for each.

physical map

political map

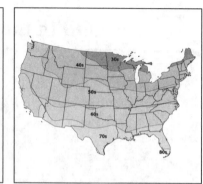

thematic map

1. You want to plan a road trip. You need to know the distances between cities and which borders you will need to cross.

   Type of map: _____

   Why would you choose this type of map?

   _____

   _____

2. You are making a presentation for school, and you want a good visual aid to show which countries are the wealthiest.

   Type of map: _____

   Why would you choose this type of map?

   _____

   _____

3. You are a city planner who is working on flood preparations. You need to know which areas of the city have the lowest elevations and are close to bodies of water.

   Type of map: _____

   Why would you choose this type of map?

   _____

   _____

Name: _____ **Date:** _____

Directions: This is a map of Africa. Use the map and scale to answer the questions.

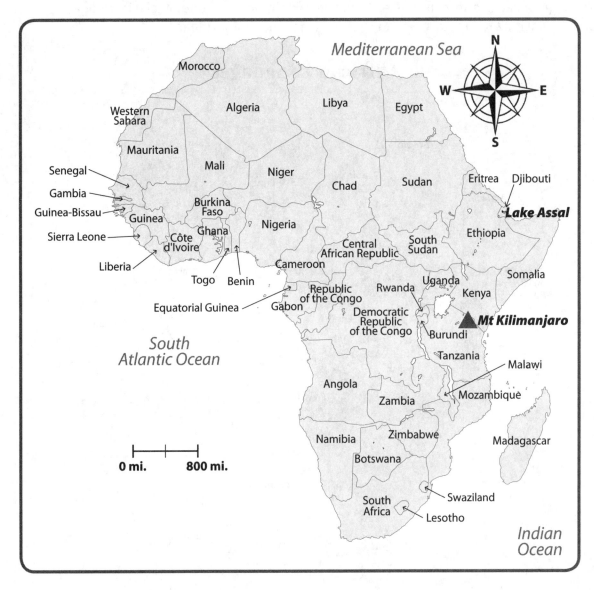

1. What political information is provided on this map of Africa?

_____

2. Outline the borders of the landlocked countries (countries that do not border the ocean).

3. Draw a line a between the highest point, Mt. Kilimanjaro, and the lowest point, Lake Assal.

4. Using the scale, what is the approximate distance between the two points?

_____

Creating Maps

Name: _____ **Date:** _____

Directions: Africa has many languages and dialects. Color the map to show the official language of each country in Africa. Countries with more than one official language are identified with 2+. Create a legend to show what your colors mean.

African Languages

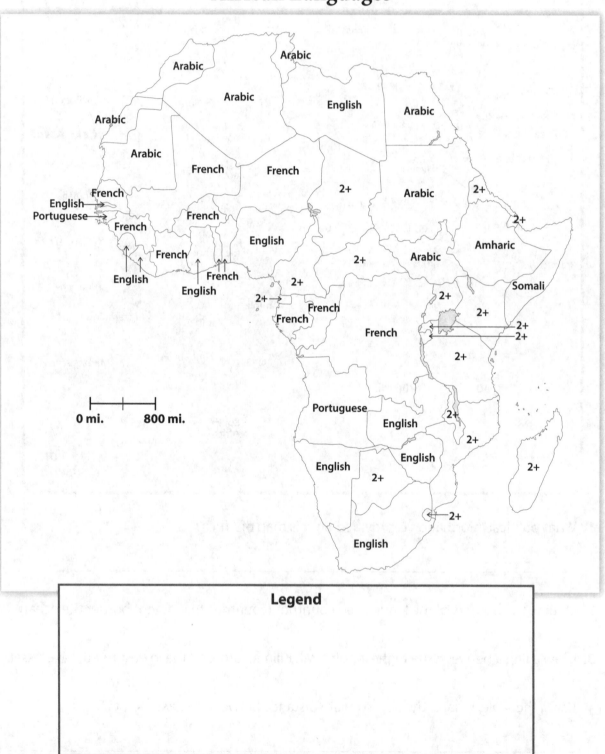

Legend

Name: _____ **Date:** _____

Directions: Read the text, and study the photo. Then, answer the questions.

Africa: The Real Melting Pot

Movies, television, and even the news sometimes falsely describe Africa. They refer to the continent of Africa as if it were a single country. Not every part of the continent is the same. These sources sometimes stereotype all of Africa as a region that is underdeveloped.

In reality, Africa is many things. There are isolated, remote areas. There are also modern cities. Africa is home to 54 countries. People speak more than 1,000 languages. There are hundreds of ethnic groups. Each group has its own history and traditions.

African countries are among the most diverse in the world. In Chad alone, there are more than 100 ethnic groups. Togo is a country smaller than the state of West Virginia. Yet in Togo, people speak 39 different languages. These differences make the culture of each African country special. They also create challenges. Schools can be hard to run because there are so many languages. Wars have been sparked by ethnic differences.

1. What do movies, television, and the news often falsely describe about Africa?

_____

_____

_____

2. What types of diversity are there in Africa?

_____

_____

_____

3. What are some of the challenges of having a very diverse country?

_____

_____

_____

Think About It

Name: _____ Date: _____

Directions: This table shows the percentage of people living in urban areas in several African countries. Study the table, and answer the questions.

Percentage of Population in Urban Areas		
Country	2012	2015
Algeria	68.9	70.7
Burundi	11.2	12.1
Chad	22.1	22.5
Djibouti	77.1	77.3
Egypt	43.0	43.1
Ethiopia	18.2	19.5
Gabon	86.4	87.2
Kenya	24.4	25.6
Libya	78.0	78.6
Madagascar	33.2	35.1
Morocco	58.7	60.2
Niger	18.0	18.7
South Africa	63.3	64.8
Somalia	38.2	39.6

1. What has happened to the urban population since 2012?

   _____

   _____

2. Which three countries have the highest percentage of people living in urban areas?

   _____

   _____

3. Why do you think some countries have small populations living in urban areas?

   _____

   _____

   _____

Name: _____ **Date:** _____

Directions: Diversity in Africa takes many forms: ethnicity, language, religion, culture, and urban areas. Think about the kinds of diversity in your community. List each type, and explain how it can have a positive effect.

_____

_____

_____

_____

_____

_____

_____

_____

_____

_____

_____

_____

_____

_____

_____

_____

_____

_____

_____

_____

_____

_____

Geography and Me

Reading Maps

Name: _____ Date: _____

Directions: Study the map, and answer the questions.

Ancient Egypt

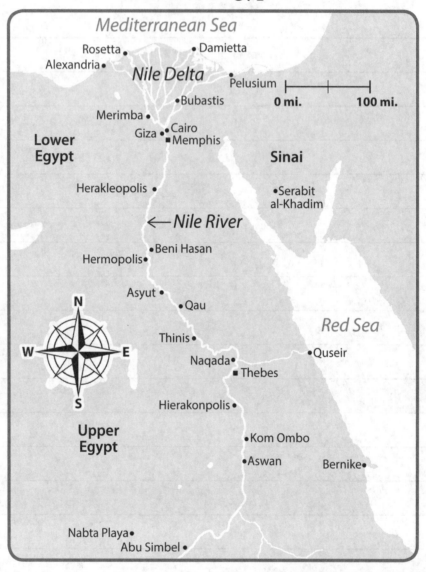

1. In what direction does the Nile River flow, and into what body of water does it empty?

_____

_____

2. Trace the Nile River on the map.

3. Use the scale to estimate the length of the Nile River on the map.

_____

Name: _____ Date: _____

Directions: Examine the pictures of ancient Egyptian monuments. Then, draw a symbol to represent each monument in the correct location on the map.

Ancient Egypt

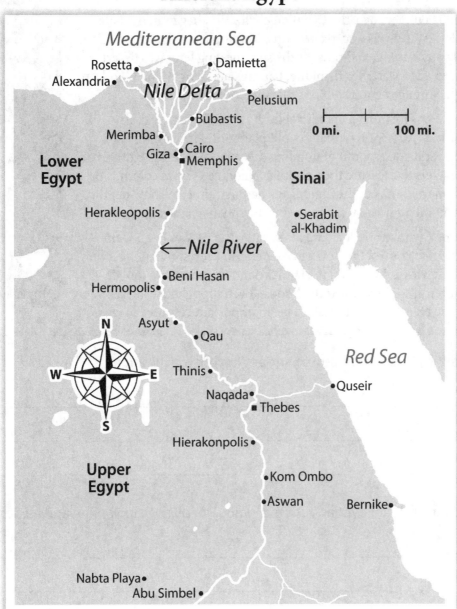

Mediterranean Sea

Rosetta •   • Damietta
Alexandria •
Nile Delta
• Pelusium

• Bubastis

Merimba •
Giza • • Cairo
■ Memphis

Lower Egypt

Sinai

Herakleopolis •

•Serabit al-Khadim

←*Nile River*

•Beni Hasan
Hermopolis•

Asyut •

• Qau

Red Sea

Thinis •

•Quseir

Naqada•
■ Thebes

Hierakonpolis •

Upper Egypt

•Kom Ombo

•Aswan   Bernike•

Nabta Playa•
Abu Simbel •

0 mi.   100 mi.

N
W   E
S

pyramids (Giza)

sphinx (Giza)

Karnak
(near Thebes)

Luxor
(near Thebes)

Read About It

Name: _____ Date: _____

Directions: Read the text, and study the photo. Then, answer the questions.

Flooding on the Nile

About 5,000 years ago, the Egyptian civilization began in the Nile River Valley. Water was a scarce resource in this part of the world. There was hardly any rainfall, and most of the surrounding land was desert. The Nile provided water for drinking, bathing, and irrigation. Floods in the Nile also brought good soil for growing food.

The Nile was fortunately predictable. It flooded every year at the same time. Ancient Egyptians could plan for their planting and harvesting. Still, the amount of flooding did vary. Ancient Egyptians used nilometers to measure the flooding. Nilometers were columns or staircases on the banks of the Nile. They had markings to show depth. When the flooding was lower than usual, Egyptians made adjustments.

In 1970, Egyptians built a dam at Aswan. The Aswan High Dam allowed them to control the flow of water and to keep some of the river's power as electricity. However, the land around the Nile is no longer as fertile after the dam stopped the floods. Even though its role has changed, the Nile River continues to be an important resource for Egypt.

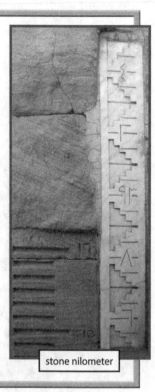
stone nilometer

1. Why was the Nile so important to ancient Egyptian civilization?

_____

_____

_____

2. How did the ancient Egyptians use technology to plan their planting and harvesting?

_____

_____

_____

3. What was the effect of building the Aswan High Dam?

_____

_____

_____

Name: _____ Date: _____

Directions: This chart shows the amount of water held in the reservoir of some of the largest dams in the world. Study the chart, and answer the questions.

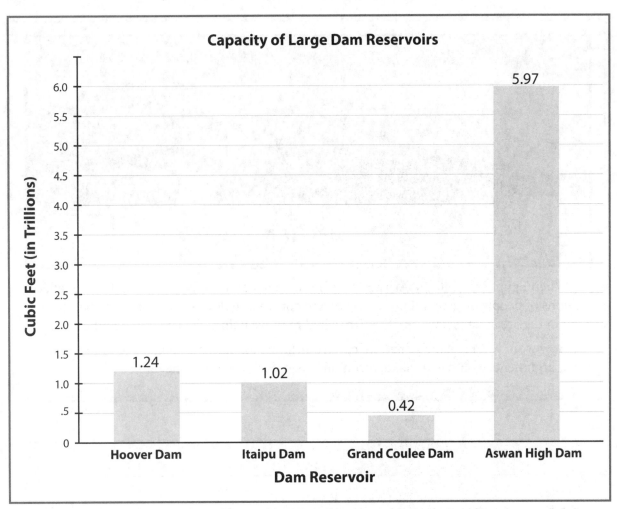

1. How much more water does the Aswan High Dam reservoir hold than the Hoover Dam reservoir?

   _____

2. Which dam do you believe generates the most electricity? Why?

   _____

   _____

3. Why do you think the Aswan High Dam reservoir holds so much water?

   _____

   _____

Geography and Me

Name: _____ Date: _____

Directions: Study the photo, and read the text. Then, use the space below to answer the questions.

When Egyptians built the Aswan High Dam, it created Lake Nasser. The dam benefits millions of people by generating electricity and controlling floods. But over 50,000 people lived in the area that is now the lake. They had to move before the lake flooded the area.

How would you feel if you had to leave your home for something like a dam to be built? What arguments would you make to convince people not to build the dam?

_____

_____

_____

_____

_____

_____

_____

_____

_____

_____

_____

Name: _____ **Date:** _____

Directions: This is a map of two regions in Northern Africa. The Sahel is the area in between the Sahara Desert and the African rainforests. Study the map, and answer the questions.

The Sahara Desert and Sahel

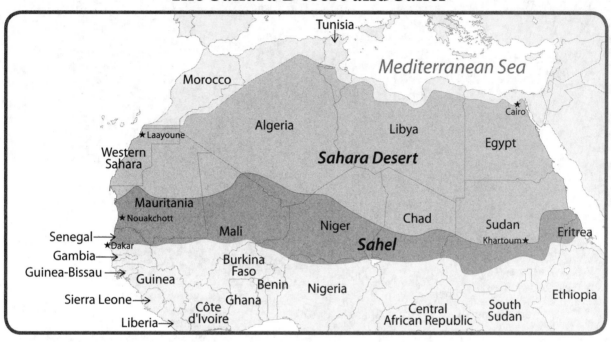

1. Most countries are not completely covered by the Sahara Desert. Which countries have the majority of their land covered by the Sahara Desert?

   _____

   _____

2. Which countries contain three or more regions, including the Sahara Desert and part of the Sahel?

   _____

   _____

   _____

3. How many major cities are shown within the Sahara Desert? List the cities.

   _____

   _____

Creating Maps

Name: _____ Date: _____

Directions: Timbuktu was a major trading center in Medieval times. Follow the steps to complete the map.

The Sahara Desert and Sahel

1. Draw a route from Cairo to Timbuktu.

2. Draw a route from Tripoli to Timbuktu.

3. Draw a route from Tunis to Timbuktu.

4. Draw a route from Ceuta to Timbuktu.

5. Use the scale to label the distance of each trade route.

Challenge: Cover the names of the countries that contain part of the Sahara Desert with sticky notes. Write the names of as many countries as you can. Then, check your answers.

Name: _____ Date: _____

Directions: Read the text, and study the photo. Then, answer the questions.

The Changing Sahara

The Sahara Desert is the largest desert in the world. It is a huge stretch of sand and rocks, with little to no rain and few sources of water. However, 5,000 to 11,000 years ago, this part of the world had plants, animals, and plenty of water. This period was called the "Green Sahara."

Scientists discovered this by studying the remains of leaf wax, shells, and dust at the bottom of the sea. They have also found skeletons of animals, such as giraffes, crocodiles, and hippos, in the Sahara. These animals could only have lived in areas with water.

cave painting in the Sahara Desert in Algeria

Other clues include human burial sites, pottery, tools, and cave paintings like the one on the right. Scientists can estimate when these humans lived and what kind of diet and lifestyle they had. These cave paintings show animals who could not have lived in a desert climate.

The climate of the Sahara has changed dramatically over time. The clues left behind show us how much climate shapes the lives of the humans and animals who live there.

1. What was the Sahara like 5,000 to 11,000 years ago?

_____

_____

_____

2. Why do you think this period was called the "Green Sahara"?

_____

_____

3. Why are cave paintings like the one above significant?

_____

_____

Think About It

Name: _____ Date: _____

Directions: Study the picture, and answer the questions.

1. How does the geography of a desert affect transportation and trade in the area?

   _____

   _____

   _____

2. How does the geography of a desert affect how people live?

   _____

   _____

   _____

3. Would it be easier for people to adapt to living in a desert that changed to a forest or in a forest that changed into a desert? Explain your answer.

   _____

   _____

   _____

   _____

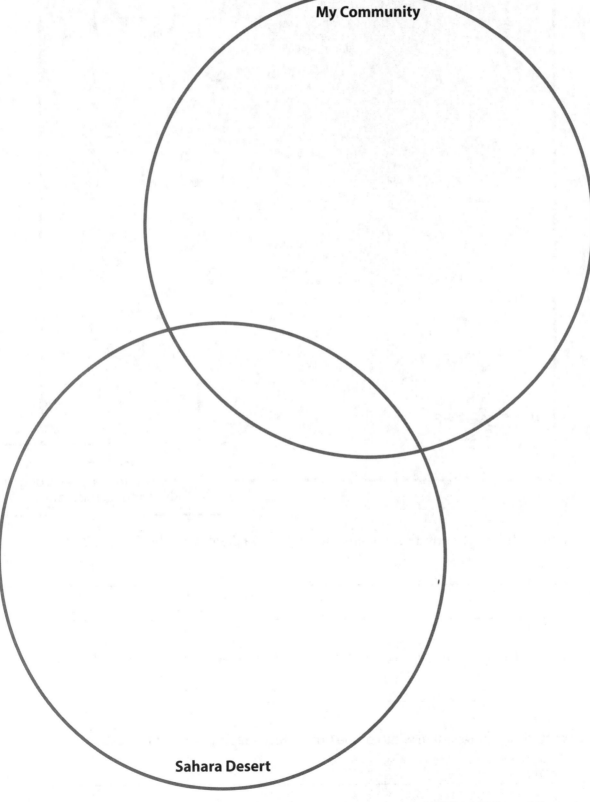

Name: _____ Date: _____

Directions: Complete the Venn diagram to compare and contrast your community with the Sahara Desert. You may consider anything, from landforms to population and lifestyle.

My Community

Sahara Desert

Name: _____   Date: _____

Directions: Study the map, and answer the questions.

Congo River Basin

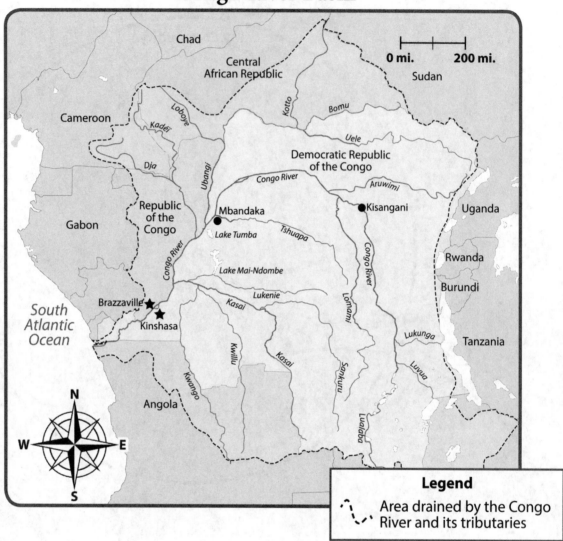

1. Name at least five countries that are part of the area drained by the Congo River.

_____

_____

2. What is the approximate length of the Congo River Basin from north to south?

_____

3. What is the approximate width of the Congo River Basin from east to west?

_____

Name: _____ Date: _____

Directions: Follow the steps to show the system of smaller rivers or streams, known as tributaries, that feed into the main river.

Congo River Basin

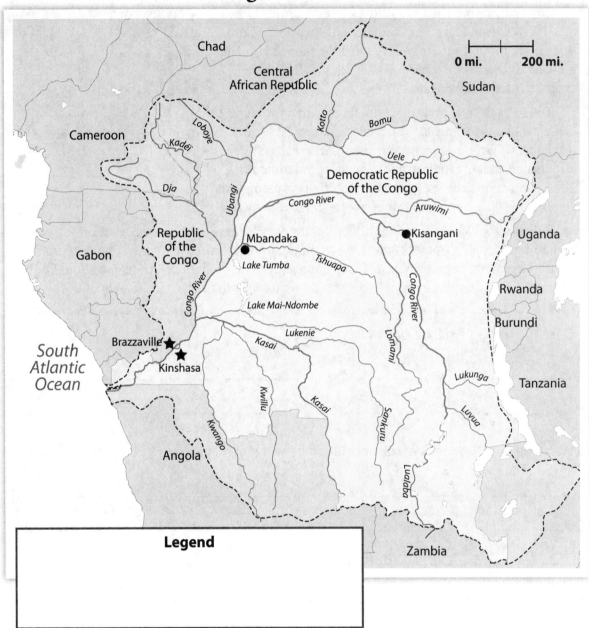

1. Trace the tributaries in one color.

2. Trace the main river in another color.

3. Create a legend to show what your colors represent.

Read About It

Name: _____ **Date:** _____

Directions: Read the text, and study the photo. Then, answer the questions.

Getting around in the Congo River Basin

The area around the Congo River is a thick forest. In fact, it's the second largest rainforest in the world. It's more than a half million square miles (800,000 square km). That is even larger than the entire state of Alaska. This huge expanse is hard to navigate.

The Congo River and the dense forest surrounding it cause problems for travel. Much of the rainforest is only reachable by river or stream. Yet many large waterfalls are impassable, making long-distance water travel difficult. The Livingstone Falls are a series of waterfalls covering over 200 miles (300 km) of river. They are too large for any boat.

Rail and road construction is difficult or impossible through the thick rainforest. Sometimes, supply shipments must use a combination of trains, trucks, and boats to make it all the way inland from the coast. Inside large cities, there is modern transportation. In the interior of the forest, most people get around on foot, by bicycle, or in small boats.

1. Describe the land in the Congo River Basin.

   _____

   _____

2. What are the challenges of transportation by water and land in the Congo River Basin?

   _____

   _____

   _____

3. How would a better transportation system benefit the countries in the Congo River Basin?

   _____

   _____

   _____

Name: _____ Date: _____

Directions: This graph shows the average monthly rainfall and temperatures in the Democratic Republic of the Congo. Study the graph, and answer the questions.

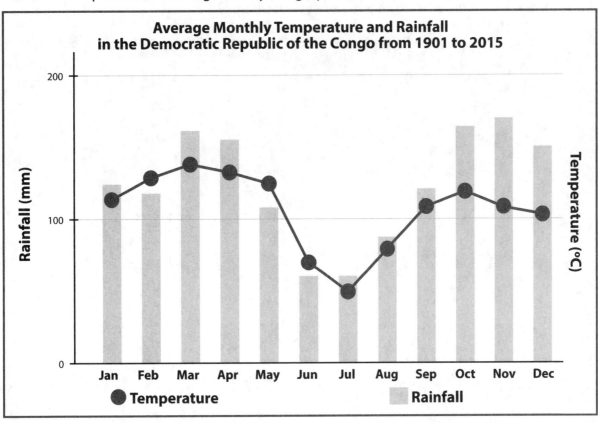

Average Monthly Temperature and Rainfall in the Democratic Republic of the Congo from 1901 to 2015

1. What is the relationship between temperature and rainfall?

_____

_____

2. Describe the rainy and dry season patterns.

_____

_____

3. When would be the best time to visit the Democratic Republic of Congo, and why?

_____

_____

_____

Name: _____ **Date:** _____

Directions: In the Congo River Basin, people often have to use multiple methods of transportation. Draw a route you might take in your community that uses multiple forms of transportation. Show what form you use at each point.

Name: _____ Date: _____

Directions: This is a map of the Swahili Coast of Africa and its trading partners in the Indian Ocean. East African trade was booming around AD 1000. Study the map, and answer the questions.

East African Trade

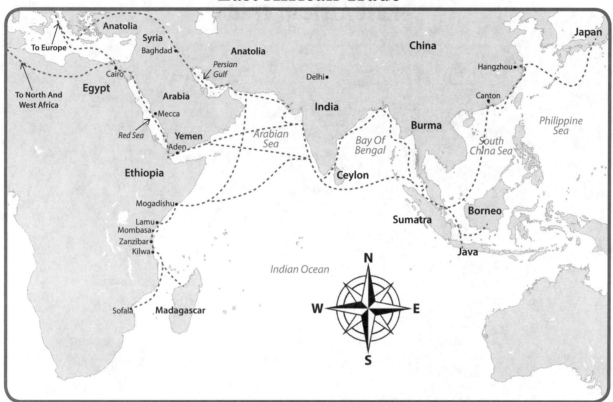

1. List at least four key trading cities along the African coast in the Indian Ocean area.

   _____

   _____

2. List at least four bodies of water used for trade on this map.

   _____

   _____

3. How many continents did these trade routes connect as shown on this map?

   _____

Name: _____ Date: _____

Creating Maps

Directions: The list below shows items traded to and from the Swahili Coast. Add symbols to the map to show where the items originated. Then, create a legend to show what your symbols represent.

East African Trade

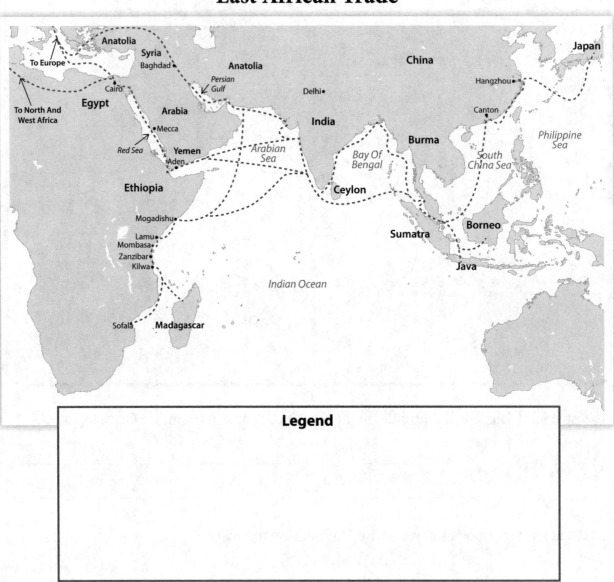

Legend

From Africa: salt, ivory, rhinoceros horns, tortoise shell, coconut oil, and gold (southern Africa)

From the Middle East: rugs

From India: jewelry, textiles

From China: porcelain

© Shell Education

Name: _____ Date: _____

Directions: Read the text, and study the photo. Then, answer the questions.

A Trading Ocean

The Indian Ocean has land on three sides. The east coast of Africa lies to the west. The Arabian Peninsula lies to the north. Asia and Australia lie to the east. It is a huge area. Yet you can travel from southern Africa to China while staying close to a coastline. This made the Indian Ocean perfect for trade. Traders were making regular trips by AD 600.

Why make the trip? Each region had different resources. Africa had gold, ivory, and rhinoceros horns. Asia and Arabia had rugs, cotton, silk, metals, and porcelain. Seasonal monsoon winds made the trip even easier.

Towns along the African coast prospered. They also became multicultural. Some of the Arabian and Asian traders stayed permanently. These foreigners had an effect on the culture and language of the area. The native Swahili culture borrowed from Arab and Persian art and architecture. The native language, Kiswahili, adopted many Persian and Arabic words. Islam came to this area from Arabia in the 9th century. It is still the dominant religion there. None of this would have happened without the use of the Indian Ocean.

1. Which lands were important trading areas along the Indian Ocean?

_____

_____

2. Why did trade develop in this area?

_____

_____

3. How did foreign cultures affect the native Swahili culture?

_____

_____

_____

Think About It

Name: _____ Date: _____

Directions: This chart shows the origins of some Swahili words. Swahili is an African language. It borrows several words from Arabic and Persian. These were two of the earliest trading partners of the Swahili coast. Use the chart to answer the questions.

Swahili Word	Meaning	Borrowed From
achari	pickle	Persian
chai	tea	Persian
katani	cotton	Arabic
rafiki	companion, friend	Arabic
safari	travel, trip	Arabic
swahili	of the coast	Arabic

1. Which words were borrowed from Arabic? Which words were borrowed from Persian?

   _____

   _____

2. Sort these words into two or three groups, and explain how you grouped them.

   _____

   _____

   _____

3. Who would most likely use these words, and in what context?

   _____

   _____

   _____

4. Why do you think these kinds of words were taken from the languages of African trading partners?

   _____

   _____

   _____

   _____

Name: _____ **Date:** _____

Directions: Complete the chart to list English words that might have been borrowed from other languages. Think about food, clothes, things you do for fun, and more.

Word	Definition	Borrowed From

Geography and Me

Reading Maps

Name: _____ Date: _____

Directions: Study the map of Asia, and answer the questions.

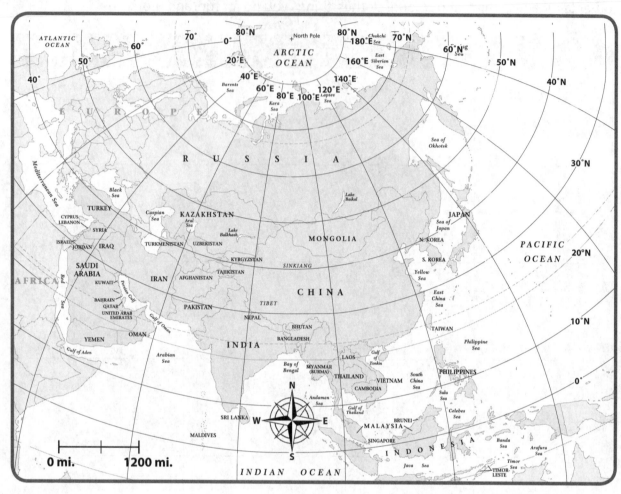

1. What are three of the largest countries in Asia based on geographic size?

_____

2. Using lines of longitude, which country is farthest to the west and which is farthest to the east?

_____

3. Using lines of latitude, how far north and south does Asia stretch?

_____

4. Which country is the longest from north to south?

_____

Name: _____ Date: _____

Directions: The box below lists countries in four regions of Asia. Use a different color to shade each region of Asia. Create a legend to identify the regions.

Regions of Asia

Central Asia: Kazakhstan, Uzbekistan, Turkmenistan, Afghanistan, Tajikistan, Kyrgyzstan, Mongolia

South Asia: Pakistan, India, Nepal, Bhutan, Bangladesh, Sri Lanka, Maldives

Southeast Asia: Myanmar, Laos, Vietnam, Thailand, Cambodia, Malaysia, Singapore, Indonesia, Brunei, Philippines, Timor-Leste

East Asia: China, North Korea, South Korea, Japan

Read About It

Name: _____ Date: _____

Directions: Read the text, and study the photos. Then, answer the questions.

The Regions of Asia

Asia is the world's largest continent. It has the most land and the most people. It is so large and diverse that it is often broken into specific regions. The most common groupings are the Middle East, Central Asia, Russia and the Caucasus, South Asia, Southeast Asia, and East Asia.

Scholars define regions in Asia by looking at their geography, culture, and history. Modern Middle Eastern countries all trace their ancestors back to this same area. Many, but not all, countries in the region have similar characteristics. Arabic is the common language, and Islam is a common religion. It makes sense to call this area a *region*.

Countries within each region also develop in similar ways. Most countries in East Asia have strong industry, big cities, and high incomes. This is not true for most countries in nearby South Asia. In these cases, labeling regions is useful. They help people describe and understand what's happening across the world.

1. How are regions in Asia defined?

_____

_____

2. What do many of the countries of the Middle East share?

_____

_____

3. Can a country still be part of a region even if it does not share all the same characteristics of other countries in the region? Why or why not?

_____

_____

Name: _____ **Date:** _____

Directions: This table shows the population in each region of Asia from 2000 to 2015. Study the table, and answer the questions.

Population by Region in Asia, 2000–2015 (thousands)				
Region	2000	2005	2010	2015
East Asia	1,496,284	1,536,540	1,575,320	1,612,287
Central Asia	55,117	58,043	62,139	67,314
South Asia	1,451,933	1,581,124	1,702,991	1,822,974
Southeast Asia	526,179	563,157	596,708	633,490
Middle East	169,019	189,753	216,390	240,459
Russia and the Caucasus	162,339	159,676	159,471	160,229

1. Describe the overall trend shown in the chart.

   _____

   _____

2. What was the overall population for Asia in 2015? How much did that increase from 2000?

   _____

   _____

3. Which region had the highest population in 2000?

   _____

4. Which region had the highest population in 2015?

   _____

5. Is the population increasing in all regions of Asia? What is one possible explanation for this?

   _____

   _____

   _____

Think About It

Name: _____ **Date:** _____

Geography and Me

Directions: Draw a map of your country. Shade different regions in your country. Draw a star where you live. Then, answer the questions.

1. With which region of your country do you most strongly identify? Why?

   _____

   _____

   _____

2. What does it mean to you to be a member of that region?

   _____

   _____

   _____

Name: _____   Date: _____

Directions: The large map shows the area known as the Fertile Crescent. This was the site of one of the earliest human civilizations. Study the maps, and answer the questions.

The Fertile Crescent

modern-day map

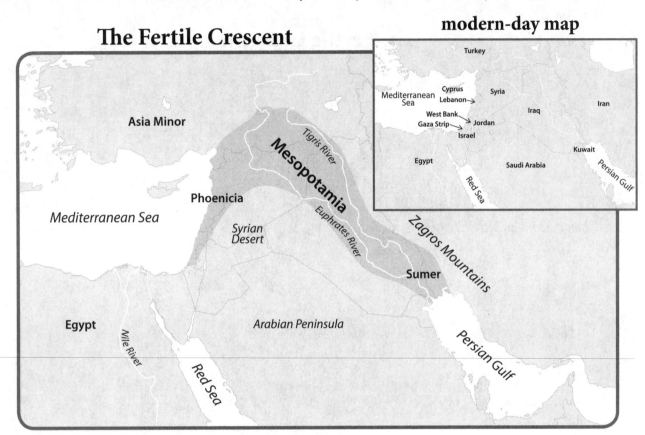

1. What physical geographic features and bodies of water are labeled on this map?

_____

_____

_____

2. List the present-day countries that make up the Fertile Crescent.

_____

_____

3. What geographic feature was most important to make this area fertile, and why?

_____

_____

Name: _____   Date: _____

Directions: Draw two land routes from the Persian Gulf to Egypt. One route should be as direct as possible. The second route should follow the rivers and the coast. Then, answer the questions.

The Fertile Crescent

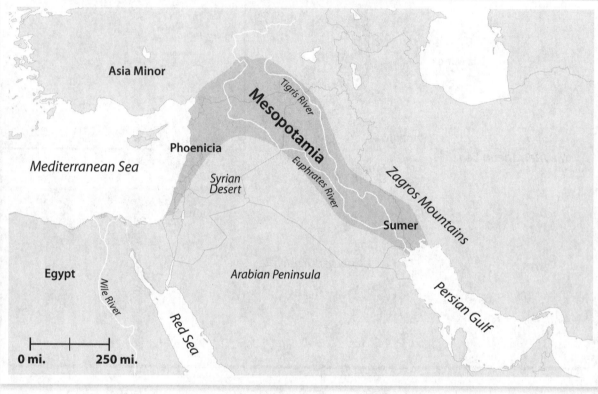

Creating Maps

1. Use the scale to estimate the length of each trip.

   _____

   _____

2. What physical features does each route pass through?

   _____

   _____

3. Which route would make the most sense for a trader, and why?

   _____

   _____

   _____

Name: _____ **Date:** _____

Directions: Read the text, and study the photo. Then, answer the questions.

Making the Fertile Crescent Fertile

All people need water to survive. This was especially true for the earliest civilizations. They had no supermarkets for food shopping and no indoor plumbing to quench their thirst. It's no surprise that all of the earliest settlements were close to rivers or other water sources.

The rise of farming made water even more important. People needed a way to get the water from the rivers to their farmland. The Tigris and Euphrates rivers in the Fertile Crescent often flooded. However, these floods were not predictable. Early farmers in the area had to plan for times of scarcity.

In Mesopotamia, in the area between the Tigris and Euphrates, farmers dug out basins next to the river to store water. They also used these basins to gather the rich soil that flowed in the water. When the fields expanded beyond the reach of these basins, Mesopotamians built canals. These canals have been rebuilt and reused over the years. Some, like the ones in the picture above, are still in use.

1. Explain two reasons early civilizations needed to be near water sources.

_____

_____

2. Why did early farmers have to store extra water?

_____

_____

3. How did Mesopotamians first get water to their fields? How did Mesopotamians improve this system?

_____

_____

_____

Think About It

Name: _____ Date: _____

Directions: This is a map of the four earliest river valley civilizations. Study the map, and answer the questions.

Early Civilizations

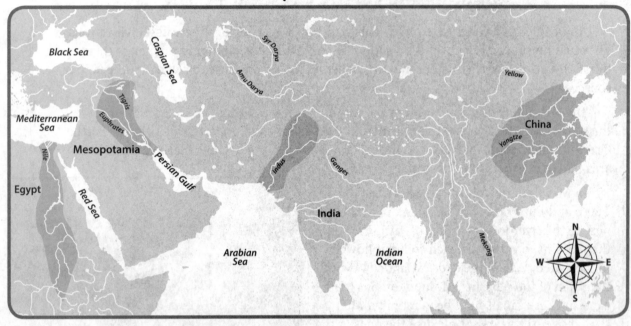

1. What part of the map has the most rivers?

   _____

   _____

2. In what parts of the modern world were these four civilizations located?

   _____

   _____

3. Based on their locations, which early civilizations would you expect to have the most interaction?

   _____

4. What likely resulted from that interaction, and why?

   _____

   _____

   _____

Name: _____ **Date:** _____

Directions: Access to clean water is still a problem in many countries. People and communities are working hard to conserve water. List things you and your community can do to conserve water.

Ways I Can Conserve Water	Ways My Community Can Conserve Water

Geography and Me

Reading Maps

Name: _____ Date: _____

Directions: Study the map, and answer the questions.

Seasonal Monsoon Winds

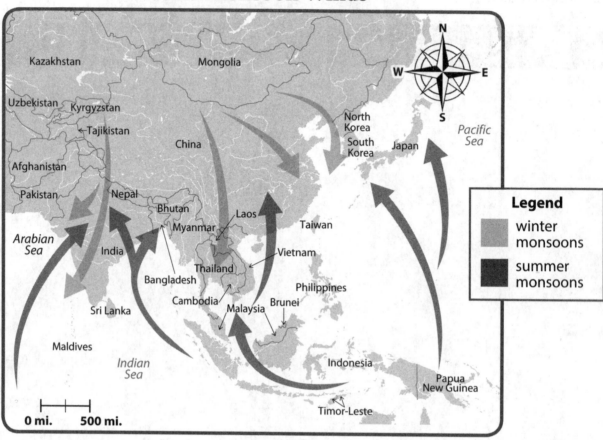

1. If you wanted to sail from Southeast Asia to India, what season would be the best time to do it, and why?

_____

_____

2. In what direction do the winter monsoon winds typically blow?

_____

3. Where is the source of the winter monsoon winds?

_____

4. Which countries would be affected first by the summer monsoon winds?

_____

Name: _____ **Date:** _____

Directions: Imagine you are a sailor living in Korea hundreds of years ago. You want to trade with as many countries as possible. But your sailboat can only travel with the help of strong winds. Follow the steps to complete the map.

Seasonal Monsoon Winds

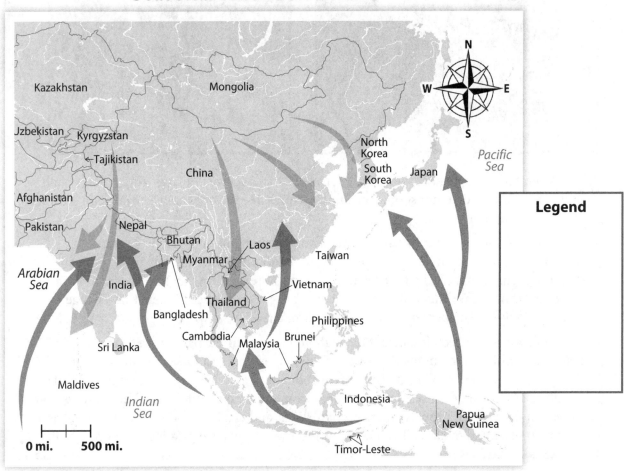

1. Draw a winter trade route.

2. Draw a summer trade route.

3. Label your routes, and add a legend to the map.

4. Explain why the routes you chose are best for each season.

_____

_____

_____

Read About It

Name: _____ Date: _____

Directions: Read the text, and study the photo. Then, answer the questions.

Monsoons

When most people hear the word "monsoon," they picture pouring rain. If you search online for pictures of a monsoon, you'll find pictures like the ones above. Monsoon season is often a wet and rainy time.

However, the real force behind the monsoon is wind, not rain. Monsoons are windstorms that blow in one direction for several months and then blow in the opposite direction. There are actually four major monsoon systems across four continents. Typically, the winds cause heavy rains and warm weather for part of the year. Then, they change direction, bringing cooler temperatures and drier air.

Monsoons have been important throughout history. In the past, trade has been both helped and hindered by monsoon winds. In monsoon areas, sailors could only travel with the wind, not against it. Changes in rainfall and temperature also affected farming. Today, scientists are learning to use energy captured from these powerful winds. Yet flooding from monsoons still causes property damage, mudslides, and deaths. Floods can spread diseases such as cholera and malaria. Even now, the monsoons still have positive and negative effects.

1. What are monsoons, and what are they sometimes associated with?

_____

2. Why are monsoon winds as important today as they have been in the past?

_____

_____

3. What are some of the problems that monsoon winds can bring?

_____

_____

Name: _____ Date: _____

Directions: This graph shows the number of malaria cases and the rainfall rate in India over time.

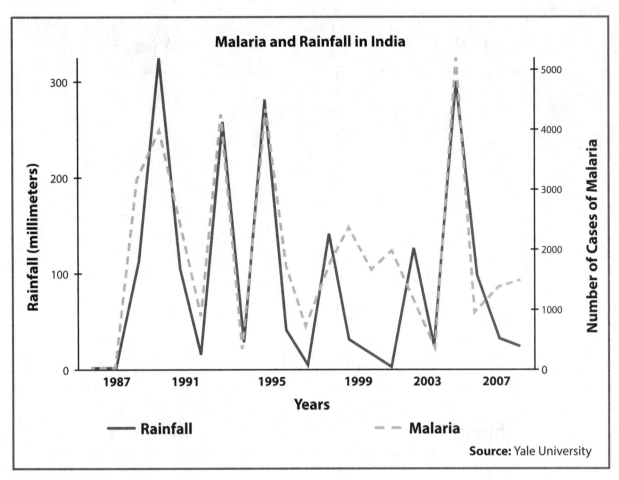

Malaria and Rainfall in India

Rainfall ———   Malaria - - -

Source: Yale University

Think About It

1. How closely do the numbers of malaria cases match the rainfall patterns?

_____

_____

2. What does this tell you about the spread of malaria?

_____

_____

3. What should the department of health in India do to combat malaria?

_____

_____

Geography and Me

Name: _____ **Date:** _____

Directions: Monsoon winds can create hot, rainy summers, and cooler, dry winters. How does that compare to the weather where you live? Describe the weather patterns in your region for each season.

Fall

Winter

Spring

Summer

Name: _____ Date: _____

Directions: This is a map of the major cities of South and Southeast Asia. Study the map, and answer the questions.

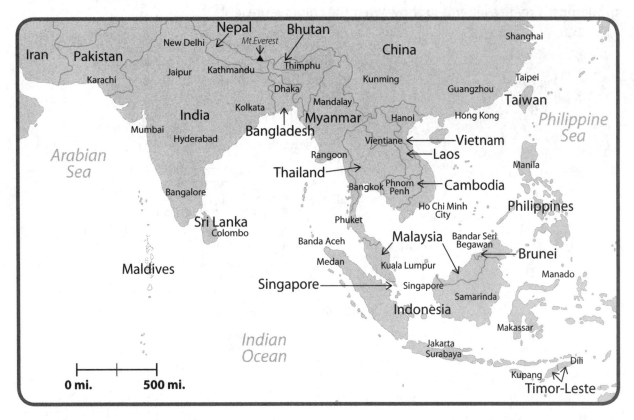

1. What is the capital of Nepal?

_____

2. How close is the capital to Mt. Everest?

_____

3. What is similar about the major cities in island nations, such as Indonesia or the Philippines?

_____

_____

4. Why might that similarity exist?

_____

_____

Name: _____ **Date:** _____

Directions: Divide the cities into three different groups based on population. An example could be 1,000,000 to 8,000,000. Then, color-code the cities on the map according to your groupings. Create a legend to show what the colors represent.

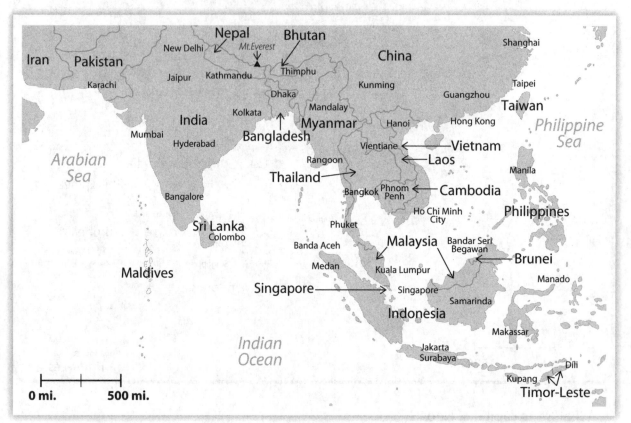

Legend

Approximate Population in 2016			
City	Population	City	Population
New Delhi	26,454,000	Kuala Lumpur	4,176,000
Mumbai	21,357,000	Ho Chi Minh City	4,389,000
Bangalore	10,456,000	Jakarta	10,483,000
Dhaka	18,237,000	Manila	9,962,000
Bangkok	6,360,000	Taipei	2,630,000

Creating Maps

Name: _____ **Date:** _____

Directions: Read the text, and study the photo. Then, answer the questions.

Population Growth in India

In 2015, the population of India was about 1.3 billion. In that same year, there were three times as many people per square mile in India than in China. That is more than 12 times higher than in the United States. Experts predict India will have the highest population in the world by 2030. That is hard to imagine in a country that is already so crowded.

It is a challenge for India to keep up with this rapid growth. Resources, such as food, clean water, sewers, and electricity, are already in short supply. So India decided to take action. The government tried to get people to have smaller families. They used stamps, posters, and even cash payments.

The actions worked. Population growth did slow in India. India's population was growing at a rate of two percent a year about 20 years ago. Today, that rate has dropped to 1.2 percent. That rate is equal to the world average. In that same period, the percentage of urban Indians living in slums decreased. Perhaps slowing down population growth gave India a chance to catch up. India is still growing. However, the pace is easier to manage.

1. Why is rapid population growth challenging for countries?

_____

_____

2. What did India do to decrease the population growth rate?

_____

_____

3. What are the pros and cons to a government trying to control population growth rate?

_____

_____

_____

Think About It

Name: _____ **Date:** _____

Directions: This table shows the percentage of people living in urban areas. Study the table, and answer the questions.

Percentage of Population in Urban Areas				
Area, Region, or Country	1950	1980	2010	2040 projected
World	29.6%	39.3%	51.6%	63.2%
High-income countries	56.6%	71.8%	79.3%	85.1%
Middle-income countries	19.0%	30.7%	48.1%	63.0%
Low-income countries	9.0%	18.4%	28.5%	43.1%
South Asia	16.0%	23.5%	32.7%	47.2%
Bangladesh	4.3%	14.9%	30.5%	50.5%
India	17.0%	23.1%	30.9%	44.8%
Pakistan	17.5%	28.1%	36.6%	52.0%
Sri Lanka	15.3%	18.8%	18.3%	25.0%
Southeast Asia	15.5%	25.5%	44.5%	60.2%
Cambodia	10.2%	9.9%	19.8%	30.6%
Indonesia	12.4%	22.1%	49.9%	67.2%
Malaysia	20.4%	42.0%	70.9%	84.2%
Philippines	27.1%	37.5%	45.3%	51.1%
Singapore	99.4%	100.0%	100.0%	100.0%
Thailand	16.5%	26.8%	44.1%	68.2%

Source: United Nations

1. Which country had the highest percentage of people living in urban areas in 2010? Which country had the lowest percentage?

_____

_____

2. What can you tell about the growth of cities (urbanization) in South Asia compared to Southeast Asia?

_____

_____

Name: _____ **Date:** _____

Directions: Imagine that your community doubled in population within one year. Describe the changes that have occurred. What is different? What is better or worse?

_____

_____

_____

_____

_____

_____

_____

_____

_____

_____

_____

_____

_____

_____

_____

_____

_____

_____

_____

_____

_____

Geography and Me

Reading Maps

Name: _____ Date: _____

Directions: Study the map, and answer the questions.

Religions in Asia

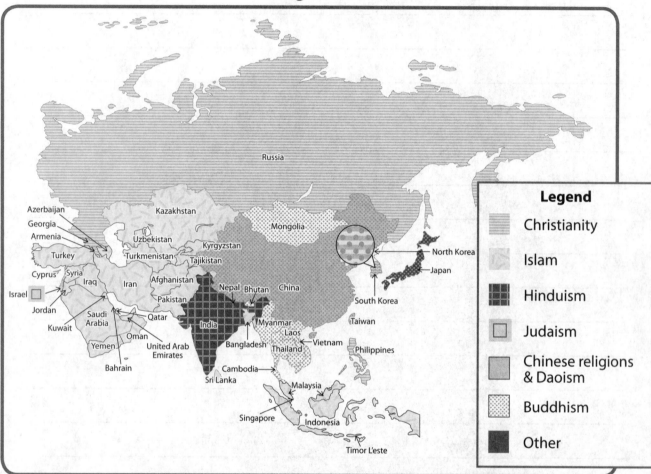

1. Which is the largest Asian country to practice Hinduism?

_____

2. Name two countries that practice more then one major religion.

_____

_____

3. What conclusions can you draw about this region of the world based on the map?

_____

_____

Name: _____ Date: _____

Directions: The major religions of the world spread to every continent. Read the text, and draw arrows on the map to show the early spread of three world religions.

Christianity: from Israel west through Turkey to Europe, and into North Africa

Islam: from Saudi Arabia throughout the Middle East, into Central Asia and North Africa

Judaism: from Israel southwest into North Africa and Ethiopia, and northwest into Eastern Europe

Read About It

Name: _____ Date: _____

Directions: Read the text, and answer the questions.

Spread of Religion

Many world religions started in Asia. Three religions began in the same area. They also share some characteristics. Those religions are Christianity, Judaism, and Islam. Jerusalem is a holy city to all three. The city is the birthplace of these three major religions. All three are monotheistic. That means they believe in one God.

Each religion spread in a different way. Judaism is the oldest of the three. Judaism has spread throughout history when its followers have migrated. In many cases, these were mass migrations due to persecution. Jewish people do not send missionaries and do not try to convert. The growth in Judaism is largely due to new births.

Christianity is a missionary religion. That means Christians try to convert others to Christianity. Much of the spread of Christianity was planned. In some cases, people were forced to convert to Christianity. One example is the Spanish arrival in the Americas. Christianity also spread when Christian empires conquered new lands.

Islam also spread through conquest and missionary activity. However, the greatest force for the growth of religions is new births. For this reason, Islam is predicted to grow much faster than other religions. By the year 2050, Islam will likely become the most popular world religion.

1. What are some things Christianity, Judaism, and Islam have in common?

_____

_____

_____

2. How did Judaism spread?

_____

_____

_____

3. How did Christianity and Islam spread?

_____

_____

_____

Name: _____ **Date:** _____

Directions: This chart shows the projected growth of religious groups in Asia and the Pacific. Study the chart, and answer the questions.

Religious Groups in Asia				
Religion	**2010 Estimated Population**	**% in 2010**	**2050 Projected Population**	**% in 2050**
Hinduism	1,024,000,000	26%	1,369,000,000	28%
Islam	986,000,000	24%	1,457,000,000	30%
*Unaffiliated	858,000,000	21%	837,000,000	17%
Buddhism	481,000,000	12%	475,000,000	10%
Folk Religions	364,000,000	9%	367,000,000	7%
Christianity	287,000,000	7%	381,000,000	7%
Other Religions	51,000,000	1%	49,000,000	1%
Judaism	200,000	Less than 1%	240,000	Less than 1%

*Unaffiliated** means not officially attached to any religion.

1. Which religion had the largest percentage of followers in 2010? Which one will have the largest percentage in 2050?

   _____

   _____

2. What might explain the predicted changes in these religions?

   _____

   _____

   _____

3. How does this chart reflect the diversity of religions in Asia?

   _____

   _____

   _____

Think About It

Name: _____ **Date:** _____

Directions: What religions are represented in your community? Create a map of your city or town with places of worship for different religions, such as, churches, temples, or mosques.

Geography and Me

Name: _____ Date: _____

Directions: This is a map of Oceania. Study the map, and answer the questions.

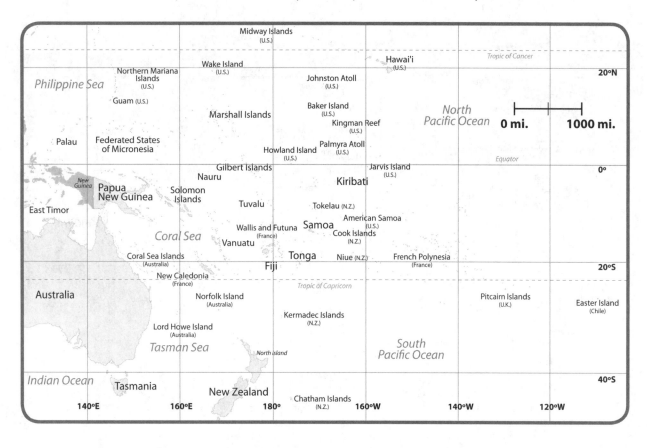

Reading Maps

1. What areas of Oceania are owned or protected by the United States?

_____

2. Using the scale, approximately how long are the two main islands of New Zealand?

_____

3. Find the approximate latitude and longitude of the following islands.

Fiji: _____

Guam: _____

Easter Island: _____

Creating Maps

Name: _____ Date: _____

Directions: Follow the steps to complete the map of Oceania and Australia.

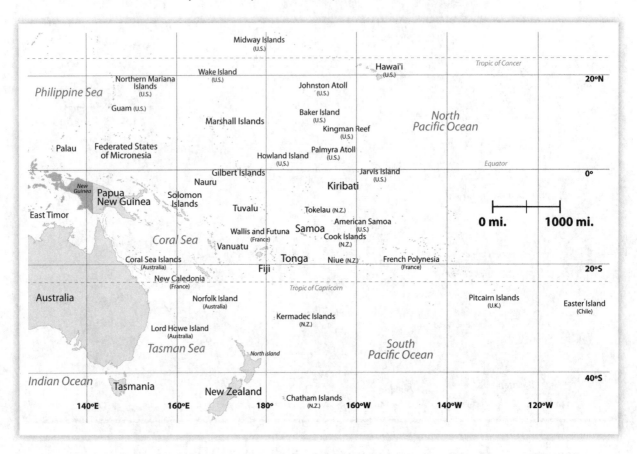

1. Draw a triangle using New Zealand, the Hawai'ian Islands, and Easter Island as the points of the triangle.

2. Shade and label the triangle *Polynesia*.

3. Draw a rectangle using Gilbert Island, Wake Island, the Marianas Islands, and Palau as the points of the rectangle.

4. Shade and label the rectangle *Micronesia*.

5. Draw a rectangle using Fiji, New Caledonia, the Solomon Islands, and New Guinea as the points of the rectangle.

6. Shade and label the rectangle *Melanesia*.

Name: _____ **Date:** _____

Directions: Read the text, and study the photo. Then, answer the questions.

Reaching the Islands

Oceania is unlike any other part of the world. There is no other inhabited area with so many people living on such small islands. The people from Southeast Asia began settling in the islands of Oceania around 1500 BC. Many islands were bare. Polynesian settlers had to bring plants and animals with them to survive.

James Cook was a European explorer. When he first came to these islands, there was much confusion. Cook could not understand where the people had come from. How had they traveled so far with only canoes? How had they made the trip with no compass? There were few written records to help answer these questions.

One reason for the confusion was Cook's lack of understanding of the monsoon winds in the area. But even into the 20th century, there were still questions. Researchers still didn't know enough about Polynesian boats and systems of navigation. To settle the matter, scholars rebuilt a Polynesian double canoe. Then, they followed directions given to James Cook by the Polynesians. They were successful. This proved that the Polynesians had been great sailors long before Europeans arrived.

1. From where and when did the first settlers come to the islands of Oceania?

   _____

   _____

2. What was confusing about Polynesian culture to European explorers and modern researchers?

   _____

   _____

   _____

3. How did researchers finally settle the matter of Polynesian exploration?

   _____

   _____

Think About It

Name: _____ Date: _____

Directions: Study the photo, and answer the questions.

traditional Maori wood-carved canoes in New Zealand

1. How does this photograph represent an important part of Polynesian culture?

_____

_____

2. What are the biggest challenges ancestors of the Maori people would have faced traveling across the ocean in wood-carved canoes?

_____

_____

_____

3. Why is it important to understand and learn about the history and culture of indigenous peoples?

_____

_____

Name: _____ **Date:** _____

Directions: Imagine you live on a small island with a small population. The nearest island is over 100 miles away. How would your life be different? What would be the same? Would you prefer to live there or live where you are now? Explain your answers.

_____

_____

_____

_____

_____

_____

_____

_____

_____

_____

_____

_____

_____

_____

_____

_____

_____

_____

_____

_____

_____

Geography and Me

Name: _____ Date: _____

Reading Maps

Directions: Recently, scientists discovered that New Zealand is actually part of an underwater continent called Zealandia. Study the map, and answer the questions.

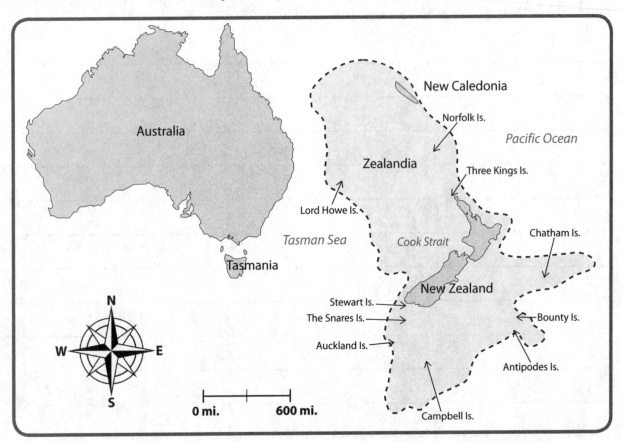

1. About how long and wide is Australia and Zealandia? Use the scale to measure each continent. Which is larger?

   _____

   _____

2. Where are most of the islands on the continent of Zealandia?

   _____

   _____

3. Approximately how close is Zealandia to Australia at its closest point?

   _____

   _____

Name: _____ **Date:** _____

Directions: Follow the steps.

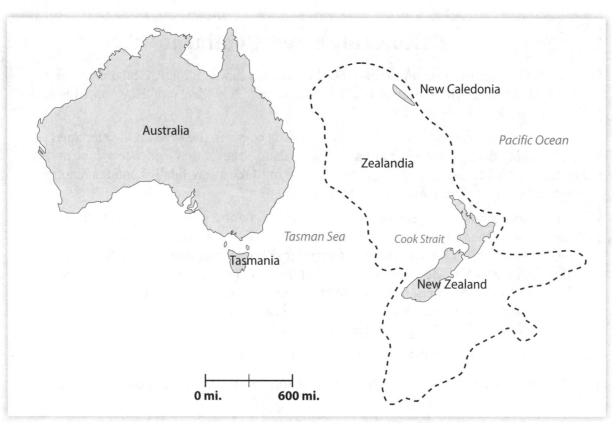

1. Use the scale to determine how large each of the two main islands of New Zealand is.

   North Island

   length: _____

   width: _____

   South Island

   length: _____

   width: _____

2. Redraw the main islands of New Zealand as many times as you need to fill the area covered by Zealandia.

3. Count how many times bigger Zealandia is than the two main islands of New Zealand.

   _____

Name: _____ Date: _____

Read About It

Directions: Read the text, and study the photo. Then, answer the questions.

Discovering a New Continent

In 2017, scientists announced that they had found a new continent, Zealandia. It is not a typical continent. The continent is a large extension of New Zealand. However, most of the land lies deep underwater.

The new area was noticed when offshore drilling began in the 1970s. Oil companies began drilling deep in the ocean around New Zealand. They found evidence of land at the bottom. Scientists collected evidence for decades until they were finally confident enough to officially declare Zealandia a separate continent.

Why does it matter? There are rules about which countries own the rights to the ocean. Whoever owns the rights can use whatever resources are there. That could be fish, seafood, or minerals. Within 12 miles (19 km) of a country's border, the resources in the sea belong to that country. If New Zealand can prove that this underwater area is actually attached, it will make a huge difference. New Zealand will gain mining and fishing rights over a vast area. That translates to a lot of money. For New Zealand, at least, this new continent could mean much more than changing maps and textbooks.

1. Is Zealandia a true continent? What is similar and what is different to other continents?

_____

_____

_____

_____

2. How was Zealandia discovered?

_____

_____

3. Why is Zealandia important for New Zealand?

_____

_____

_____

Name: _____ **Date:** _____

Directions: This map shows the continents plus the new continent of Zealandia. Study the map, and answer the questions

Earth's Tectonic Plates and Continents

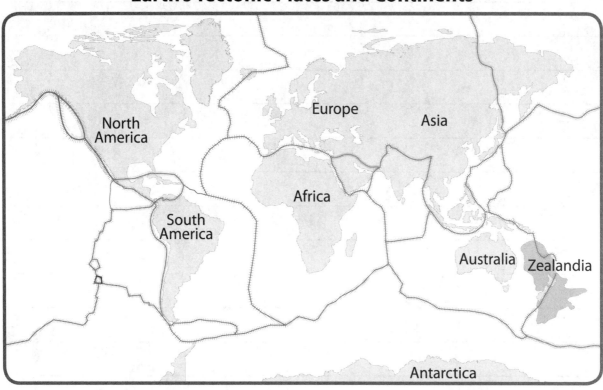

1. What do you think should be the definition of a continent?

_____

_____

2. Should Zealandia be widely accepted as an eighth continent? Why or why not?

_____

_____

_____

3. Why might countries besides New Zealand be interested in Zealandia?

_____

_____

Name: _____ **Date:** _____

Directions: Imagine that the continents had never split apart and we all lived on one giant land mass. How might history be different? How might your life today be different?

Geography and Me

_____

_____

_____

_____

_____

_____

_____

_____

_____

_____

_____

_____

_____

_____

_____

_____

_____

_____

_____

_____

_____

_____

_____

Name: _____ Date: _____

Directions: Study the map of Australia, and answer the questions.

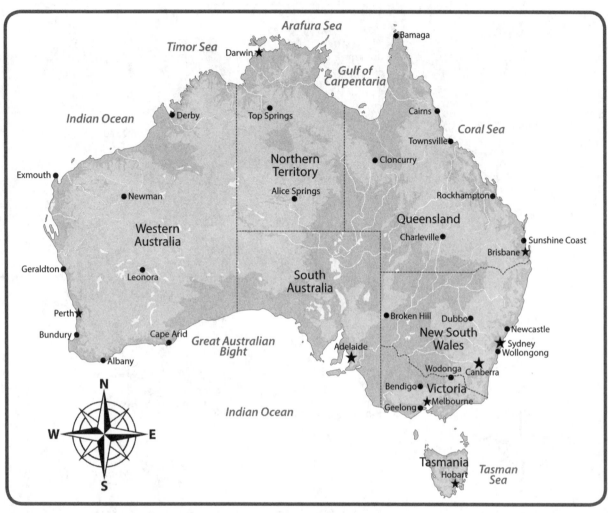

1. Australia is divided into seven states. List each state.

_____

_____

2. Which two states have the majority of Australian cities?

_____

3. What bodies of water surround Australia?

_____

_____

_____

Creating Maps

Name: _____ Date: _____

Directions: Read the text, and follow the steps.

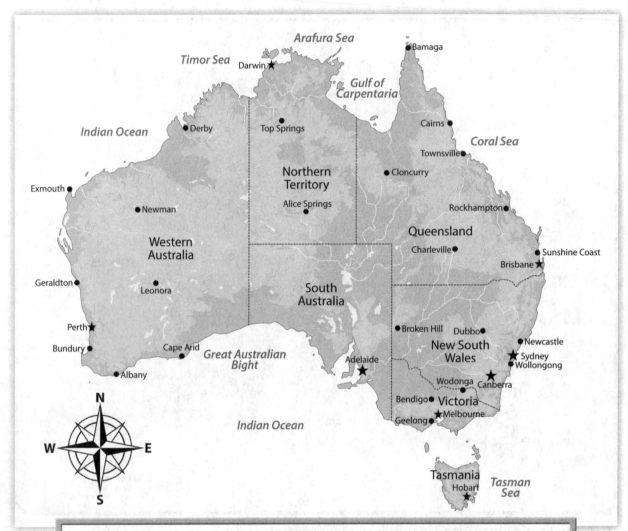

Your friends want to take a road trip across Australia. They want to stop at the following cities, in whatever order makes sense: Adelaide, Brisbane, Cairns, Darwin, and Sydney. They will be starting and finishing in Melbourne.

1. Trace a route for the road trip on the map. Do not cross any lines on the route.

2. List the directions for this route.

_____

_____

_____

_____

Name: _____ Date: _____

Directions: Read the text, and study the photo. Then, answer the questions.

The Outback

Many people picture kangaroos in the desert when they think of the Outback of Australia. That is certainly part of the Outback. The Outback covers more than 70 percent of Australia. That is more than half the size of the United States. It includes four deserts. It also has mountains, tropical grasslands, forests, scrub, and even some wetlands. What each of these lands has in common is a lack of rain.

As a dry and hot place, the Outback is not well populated. Less than 5 percent of the population of Australia lives there. Most residents of the Outback are either involved in raising livestock or mining. About one-quarter of the people living in the Outback are Aboriginals, or native Australians.

Aboriginals came to the Outback more than 30,000 years ago. They developed methods of survival and caring for the land. Aboriginals managed fires and dug waterholes. Today, government agencies are beginning to acknowledge and use this "desert knowledge" to better manage and care for the land. It is one of the few wild areas on Earth. It deserves protection.

1. How big is the Outback, and what kind of land can you find there?

   _____

   _____

2. What would be an advantage of living in the Outback?

   _____

   _____

3. Explain two strategies Aboriginals used to manage the land.

   _____

   _____

   _____

   _____

Think About It

Name: _____   Date: _____

Directions: Australia was founded by convicts that were sent from England. Study the illustration below from 1830, and answer the questions.

1. Describe the people you see in the illustration.

   _____

   _____

2. What would be some of the biggest challenges to starting a colony with convicts?

   _____

   _____

3. Why do you think Australia was chosen as a place to send English convicts?

   _____

   _____

Name: _____ **Date:** _____

Directions: For many people, Australia is a dream travel destination. But would you like to live there? Write pros and cons of moving to Australia. Be sure to consider the land and climate. Then, answer the question.

Pros (Advantages)	Cons (Disadvantages)

1. Examine your lists. Would you like to live in Australia? Why or why not?

_____

_____

_____

_____

_____

Reading Maps

Name: _____   Date: _____

Directions: Use the map of vegetation in Australia to answer the questions.

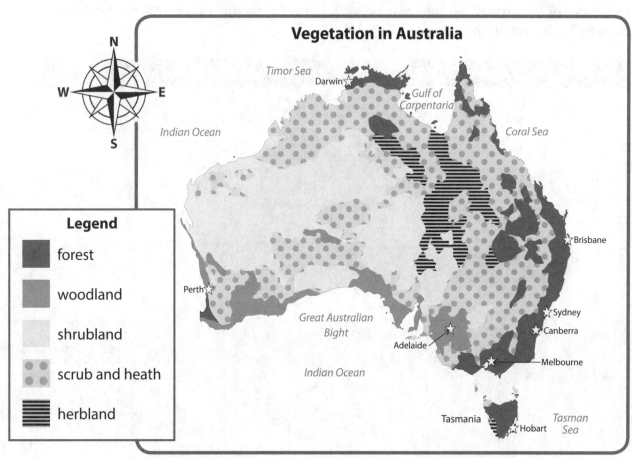

1. In what two types of vegetation do the majority of Australia's population live?

   _____

   _____

2. Describe where most of the Australian herbland is located.

   _____

   _____

3. What type of vegetation is most dominant in western Australia?

   _____

4. What type of vegetation is most dominant on Tasmania?

   _____

Name: _____ Date: _____

Directions: Use the chart to label the map. Use symbols to show where some Australian animals live. Then, create a legend to show what your symbols represent.

Animals in Australia

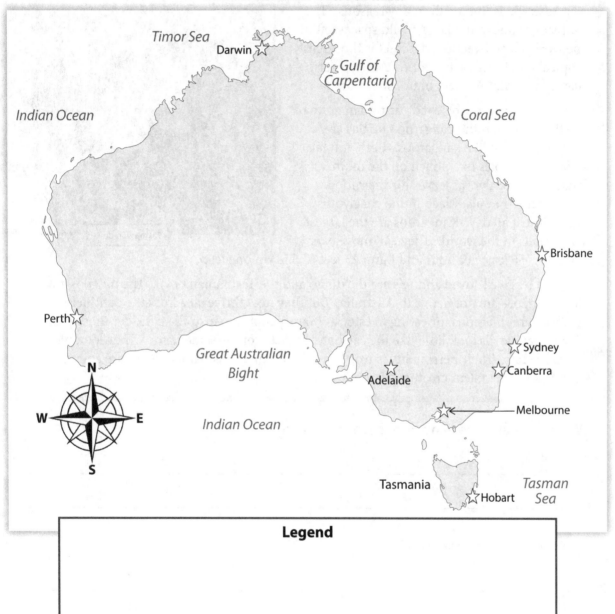

Animal	Location
koala	midway up the eastern coast
kangaroo	center of Australia and on Tasmania
Australian dingo	north, south, east, and west
Tasmanian devil	on Tasmania
platypus	along the northeast coast and on Tasmania

Name: _____ Date: _____

Read About It

Directions: Read the text, and study the photo. Then, answer the questions.

Australia's Animals

Australia is unique for many reasons. The Outback is one of the largest wild spaces left in the world. The Great Barrier Reef is the largest tropical coral reef in the world. Australia also has an incredible variety of unusual animals.

Two of the most beloved Australian animals are the koala and the kangaroo. Koalas are marsupials, just like kangaroos. Both animals carry their babies in a pouch on the mothers. Koalas use eucalyptus leaves for their food and water. They also sleep in the eucalyptus trees almost all day. Kangaroos are the largest marsupials in the world. They can move over 35 miles (56 km) per hour and jump 25 feet (7.6 m) in one leap.

Tasmanian devil

Two less well-loved animals are the dingo and the Tasmanian devil. The dingo is a wild dog. Dingoes are not native to Australia. But they now thrive all across the continent. Dingoes are infamous for eating livestock. Many people in Australia consider them pests. Tasmanian devils look little like the cartoon character of the same name. They look more like large rats. Their personality is worthy of their name. Tasmanian devils are angry, territorial, and violent creatures.

1. What do koalas and kangaroos have in common?

_____

_____

2. Why are dingoes unpopular?

_____

_____

3. How are Tasmanian devils described in the text?

_____

_____

Name: _____ Date: _____

Directions: Study the table, and answer the questions.

Numbers of Threatened Species in Australia and the World in 2016		
Type of Species	Australia	World
fish	118	8,124
mammals	63	3,406
birds	50	4,393
plants	92	15,056

Source: World Bank

1. Calculate the percentage of Australia's threatened species in the world for each category.

   _____

   _____

2. In which category does Australia have the highest number of threatened species? What might be the reason for that?

   _____

   _____

   _____

3. Which of these categories do you think gets the least attention, and why?

   _____

   _____

   _____

4. What are some steps the Australian government might take to protect these species?

   _____

   _____

   _____

   _____

Think About It

Geography and Me

Name: _____ **Date:** _____

Directions: What endangered animals do you know of in the world today? Draw or write a list in the box. Then, answer the question.

1. What is the biggest threat to any animal on the planet, and why?

_____

_____

_____

Name: _____ **Date:** _____

Directions: Study the map, and answer the questions.

Major Coral Reefs

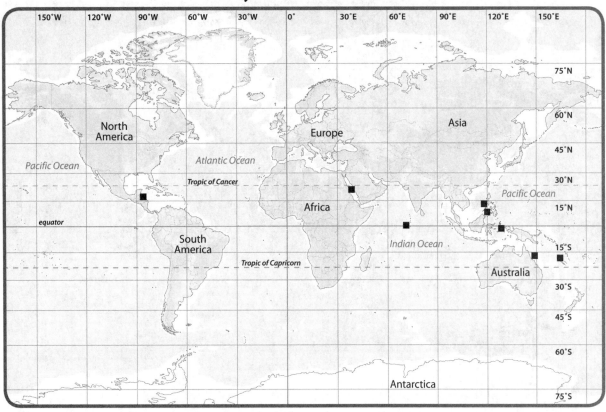

1. Which continents have major coral reefs?

_____

_____

2. In which oceans or seas are each of the coral reefs located?

_____

_____

_____

3. In what latitude range are the coral reefs located?

_____

Creating Maps

Name: _____   Date: _____

Directions: Label the coral reefs on the map.

Major Coral Reefs

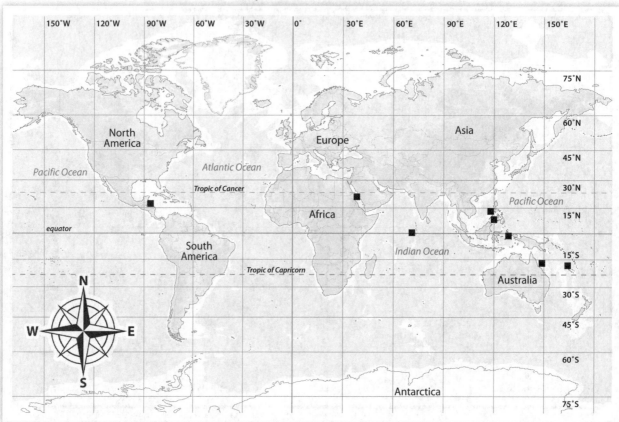

Apo Reef: northernmost reef in the Pacific Ocean

Belize Barrier Reef: Caribbean Sea (near North America)

Great Barrier Reef: Australia

Maldives Reef: in the middle of the Indian Ocean

New Caledonia Reef: near the islands east of Australia

Raja Ampat Islands Reef: Indonesia (north of Australia)

Red Sea Reef: in the Red Sea (between Africa and Asia)

Tubbataha Reef: south of the Apo Reef

Name: _____ Date: _____

Directions: Read the text, and study the photo. Then, answer the questions.

Reefs at Risk

Coral reefs are underwater ridges made of coral, seaweed, and sea sponges. They are home to 25 percent of all life in the oceans. They provide shelter, protection, and food in the soft polyps of the coral. The coral is also a living creature. That makes its preservation doubly important.

Coral reefs are in danger mostly due to human actions. Anything we do that upsets the balance of life in the reef is destructive. Overfishing can cause algae growth. The use of dynamite in fishing damages coral. Tourists cause accidental damage to coral through diving, boating, and littering.

Climate change is also damaging coral reefs. As the sea temperature rises, the coral turns white. This is called *coral bleaching*. It does not take a large increase, either. A single degree Celsius is enough if it lasts for several weeks. If the temperature stays high longer, the coral will die. When all the coral dies, the reef dies. Everything that lives in the reef will be affected.

Coral reefs grow very slowly. The Great Barrier Reef took five million years to form. This means that any destruction we cause would take a very long time to heal. There is no time to waste.

1. What are coral reefs?

_____

_____

2. How are the actions of humans damaging coral reefs?

_____

_____

3. What causes coral bleaching, and what can be the result of it?

_____

_____

Think About It

Name: _____ **Date:** _____

Directions: Coral is easily damaged by the slightest touch from scuba divers. Study the photo of the diver in a reef, and answer the questions.

1. What kind of damage could scuba divers cause to a coral reef?

_____

_____

2. What is an alternative to scuba diving as a way to enjoy and explore coral reefs?

_____

_____

3. What would you enjoy the most if you were able to scuba dive in a coral reef, and why?

_____

_____

_____

Name: _____ **Date:** _____

Directions: What steps could you and your friends take to help preserve coral reefs? List actions you could take today. Put a star next to things you are already doing.

_____

_____

_____

_____

_____

_____

_____

_____

_____

_____

_____

_____

_____

_____

_____

_____

_____

_____

_____

_____

_____

_____

Geography and Me

Reading Maps

Name: _____ **Date:** _____

Directions: This is a map of the Roman Empire at its greatest size, at around AD 117. Study the map, and answer the questions.

Roman Empire

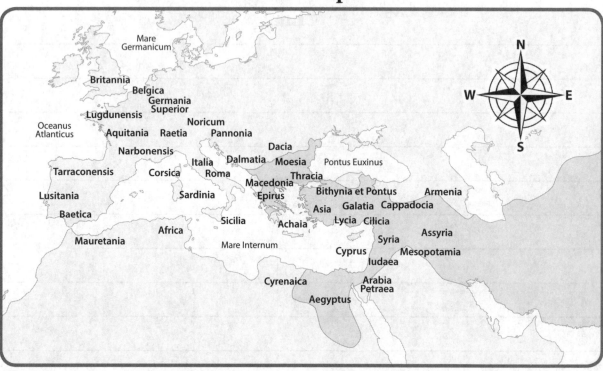

1. What are the furthest points in the empire to the east and west?

_____

_____

2. Describe the size and location of the Roman Empire.

_____

_____

_____

_____

3. This map uses the Latin names for places. Which names do you recognize as similar to the English names?

_____

_____

Name: _____   Date: _____

Directions: This map shows the ancient Greek Empire under Alexander the Great. Shade the areas that were under Roman control 400 years later. Add a legend with items: Greek Empire, Roman Empire, and shared territory. Then, answer the question.

Greek Empire

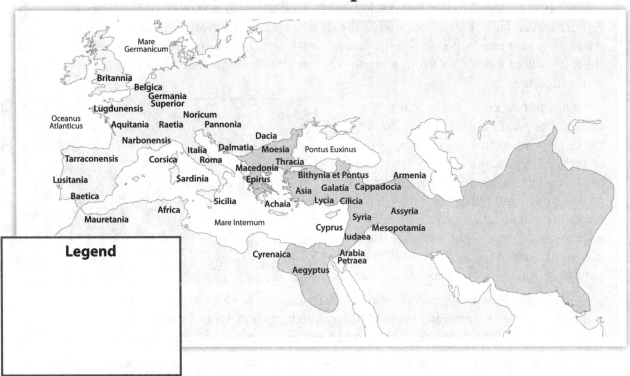

Legend

Roman Empire: Britannia, Belgica, Germania Superior, Lugdunensis, Aquitania, Narbonensis, Raetia, Noricum, Tarraconensis, Lusitania, Baetica, Corsica, Sardinia, Italia, Sicilia, Pannonia, Dalmatia, Dacia, Moesia, Thracia, Mauretania, Africa, Cyrenaica

Shared Territory: Epirus, Macedonia, Thracia, Asia, Bithynia et Pontus, Cappadocia, Armenia, Assyria, Mesopotamia, Galatia, Lycia, Cilica, Syria, Iudaea, Aegyptus

1. How might these two large empires have affected the regions they conquered?

_____

_____

_____

_____

Read About It

Name: _____   Date: _____

Directions: Read the text, and study the photo. Then, answer the questions.

Ancient Greece and Rome: A Shared Past

There are reasons why ancient Greece and Rome are paired together in history books. First, they overlapped in territory, including the land that is Greece today. The Roman Empire spread farther west and south. The Greeks conquered more land to the east in Asia. They also overlapped in time. Ancient Greek civilization began centuries before the founding of Rome. Yet Roman expansion was already underway when the Greek civilization fell.

Ancient Greece passed on culture and territory to Rome. Romans adopted the Greek gods with a few changes. For example, Zeus became Jupiter. Poseidon became Neptune. Romans built upon Greek philosophical and political ideas. Even Greek art and architecture had a heavy influence on Rome. That influence is seen in the entrance to the Pantheon, a Roman temple. It is lined with Greek-style columns.

Greeks and Romans spread their cultures across Europe through conquest. Yet even centuries after these empires crumbled, their influence continues to shape the world. People still look to ancient Greek and Roman art, architecture, and ideas for inspiration.

1. Why did ancient Greece and Rome share so much culturally?

_____

_____

2. How does the photo of the Pantheon above show the Greek influence on ancient Rome?

_____

_____

3. How were the ancient Greeks and Romans able to spread their culture across Europe?

_____

_____

_____

Name: _____ **Date:** _____

Directions: This chart shows Greek and Roman gods and goddesses. Though the names changed, most of the characteristics, stories, and responsibilities of these gods stayed the same. Study the chart, and answer the questions.

Greek Name	Roman Name	Key Responsibilities
Aphrodite	Venus	beauty and love
Ares	Mars	war
Artemis	Diana	hunting
Hades	Pluto	the underworld
Hera	Juno	marriage and family
Hermes	Mercury	travel and trade
Poseidon	Neptune	the sea
Zeus	Jupiter	the sky; the ruler of the gods

1. What do you notice about most of the Roman names for these gods?

_____

_____

2. Why do you think the largest planet shares a name with the ruler of the gods?

_____

_____

_____

3. What do the key responsibilities of these gods suggest about the values of ancient Greeks and Romans?

_____

_____

_____

Name: _____ **Date:** _____

Geography and Me

Directions: In ancient times, ideas often spread through wars and conquest. Today, there are still wars, but there is little conquest. Yet ideas spread even more easily. In what ways does the spread occur? Brainstorm a list, and explain your answers.

_____

_____

_____

_____

_____

_____

_____

_____

_____

_____

_____

_____

_____

_____

_____

_____

_____

_____

_____

_____

_____

_____

Name: _____   Date: _____

Directions: Volcanoes are marked with triangles. The plate boundary is shown with a dashed line. Study the map, and answer the questions.

Major Volcanoes in Italy

1. List the major volcanoes in Italy.

   _____

   _____

2. Which major cities in Italy are in the greatest danger of volcanoes and earthquakes?

   _____

3. What do you notice about the location of the plate boundary, the edge of the underground tectonic plates?

   _____

   _____

Creating Maps

Name: _____ Date: _____

Directions: Follow the steps to show different levels of destruction after a volcanic eruption.

Major Volcanoes in Italy

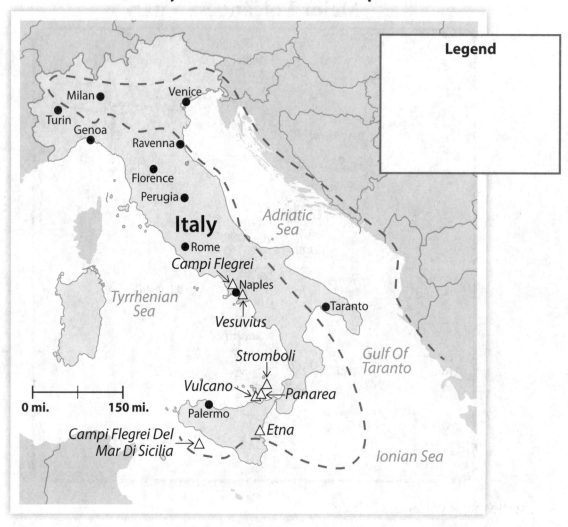

1. Draw a ring around Vesuvius and Etna. Each ring should extend 50 miles out.

2. Draw a second ring around each volcano that reaches 100 miles out.

3. Draw a third around each volcano that reaches 200 miles out.

4. Add a legend to the map. It should include the symbols for volcanoes, the plate boundary, and the levels of destruction.

Name: _____ Date: _____

Directions: Read the text, and study the photo. Then, answer the questions.

Vesuvius: Destroyer and Protector

Mount Vesuvius is an active volcano in Italy. It is infamous for destroying ancient cities. It is also known for preserving history. In AD 79, Mount Vesuvius erupted. The cities of Herculaneum, Stabiae, and Pompeii were all destroyed. But ash and rocks covered Pompeii so quickly that they sealed the town like a blanket.

More than 18 feet (5.5 meters) of debris covered the town within a few days. People who had not fled for safer ground died quickly. It was a tragedy. But the town itself was well preserved and left under the ash for years.

In the eighteenth century, archaeologists began to dig out the city. They found ancient houses, restaurants, and temples. They also found objects such as paintings and even food. The remains found in Pompeii gave us a clearer picture of everyday life in ancient Rome. Researchers in Italy are still studying the site. Meanwhile, scientists work to improve their ability to predict volcanic eruptions and prevent further disasters.

Read About It

1. What is Mount Vesuvius?

_____

_____

2. What happened in AD 79?

_____

_____

3. Why is Pompeii an important archaeological site?

_____

_____

4. What makes Pompeii so different from other sites in Italy?

_____

_____

Name: _____ Date: _____

Directions: This table shows all the volcanoes in Italy with observed eruptions.

Volcano Name	Volcano Type	Last Eruption	Elevation (meters)	Population within 30 km
Campi Flegrei	caldera	1538	458	3,006,865
Campi Flegrei Mar Sicilia	submarine	1867	-8	230,578
Etna	stratovolcano(es)	2017	3,295	1,016,540
Ischia	complex	1302	789	383,661
Larderello	explosion crater(s)	1282	500	106,985
Pantelleria	shield	1891	836	12,403
Stromboli	stratovolcano	2017	924	3,894
Vesuvius	stratovolcano	1944	1,281	3,907,941
Vulcano	stratovolcano(es)	1890	500	86,766
Vulsini	caldera	104 BC	800	195,475

Source: Global Volcanism Program, Smithsonian Institute

1. What conclusions can you draw about how active the volcanoes are in Italy?

_____

_____

2. What kinds of volcanoes are most recently active in Italy?

_____

_____

3. Why is the population information a useful part of this chart?

_____

_____

4. How might the Italian government use this information?

_____

_____

Think About It

Name: _____ **Date:** _____

Directions: Imagine that you were alive in AD 79 when Vesuvius erupted. Write a journal entry describing what you see from Pompeii.

_____

_____

_____

_____

_____

_____

_____

_____

_____

_____

_____

_____

_____

_____

_____

_____

_____

_____

_____

_____

_____

Geography and Me

Reading Maps

Name: _____ **Date:** _____

Directions: Study the map of Scandinavia, and answer the questions.

Scandinavia

1. Which country has the northernmost point in Scandinavia?

   _____

   _____

2. Notice the location of the Arctic Circle. What do you expect the climate to be like in this area?

   _____

   _____

3. What do you notice about the coastlines of these countries?

   _____

   _____

Creating Maps

Name: _____ **Date:** _____

Directions: Follow the steps to complete the map.

Scandinavia

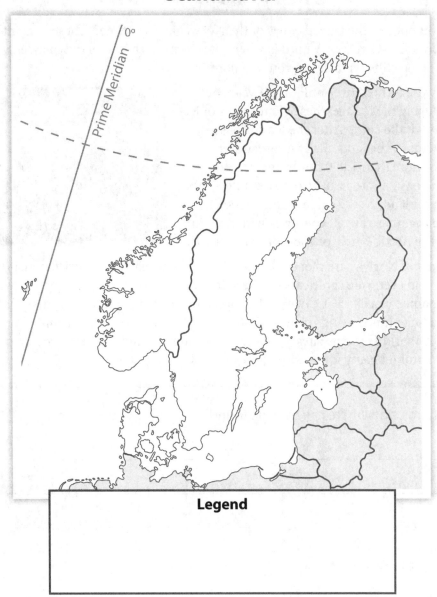

Legend

1. Color the individual countries of Scandinavia. Use a different color for each one.

2. Trace the Arctic Circle in a different color.

3. Create a legend for the map. Include the Arctic Circle and the names of the countries.

4. Label any other countries you know on the map.

Name: _____ Date: _____

Read About It

Directions: Read the text, and study the photo. Then, answer the questions.

The Fjords of Norway

Have you noticed the jagged edges of the Norwegian coastline? These cuts into the land are fjords (fee-ORDZ). A fjord is where the ocean stretches inland in a deep, narrow waterway. Steep cliffs typically surround a fjord.

Fjords were formed over millions of years by glaciers. During the ice age, these giant sheets of ice carved through the land. After the ice age passed, these deep cuts in the land filled with seawater. Norway has more than 1,000 fjords, mostly along the western coast. Fjords can also be found in Chile, Canada, Greenland, New Zealand, and Alaska. It's no surprise that these are all countries with land in or close to the Arctic Circle or the Antarctic Circle.

The fjords of Norway are plentiful, beautiful, and popular with tourists. Visitors hike, fish, kayak, and even ride cable cars up the sides of the cliffs. In fact, two of Norway's fjords have been named as UNESCO World Heritage sites. This designation means that the fjords are so valuable to humankind that the United Nations is committed to protecting them. Unlike other World Heritage sites, Norway's fjords are not under threat. People can safely enjoy their natural beauty for years to come.

1. What are fjords, and where can they be found?

_____

_____

_____

2. How were fjords formed?

_____

_____

3. Why are fjords popular tourist spots?

_____

_____

Name: _____ **Date:** _____

Directions: This is the Geiranger Fjord in Norway. Study the photo and map, and answer the questions.

a town along the Geiranger Fjord

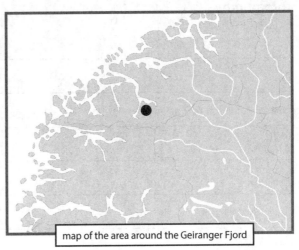

map of the area around the Geiranger Fjord

1. Describe the land surrounding the Geiranger Fjord.

_____

_____

_____

2. What might be some of the advantages of living in a town along a fjord like this?

_____

_____

_____

3. How might fjords affect the culture of an area like Geiranger?

_____

_____

4. What natural and man-made factors continue to affect fjords?

_____

_____

Think About It

Name: _____ **Date:** _____

Directions: Think about the closest bodies of water to your home. What are they? Why might you or other people visit them? Explain any useful purpose these bodies of water play in your community.

_____

_____

_____

_____

_____

_____

_____

_____

_____

_____

_____

_____

_____

_____

_____

_____

_____

_____

_____

_____

_____

Geography and Me

Name: _____ Date: _____

Directions: This is a map of Europe showing the dates when each country developed its first railway line. Study the map, and answer the questions.

Railroad Development in Europe

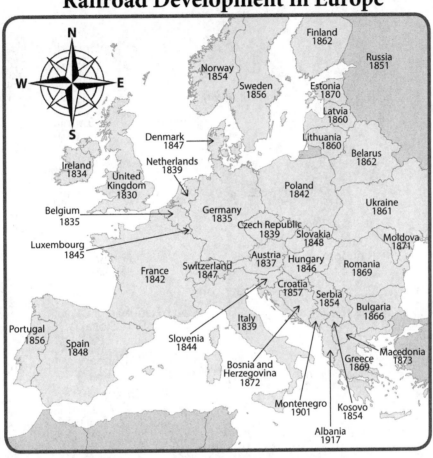

1. What direction did railways generally expand through Europe?

_____

2. How many years was it between the construction of the first railway line in Europe to the last?

_____

3. Were there any exceptions to the pattern above? If so, what are they?

_____

_____

Name: _____ Date: _____

Creating Maps

Directions: Follow the steps to complete the map.

Railroad Development in Europe

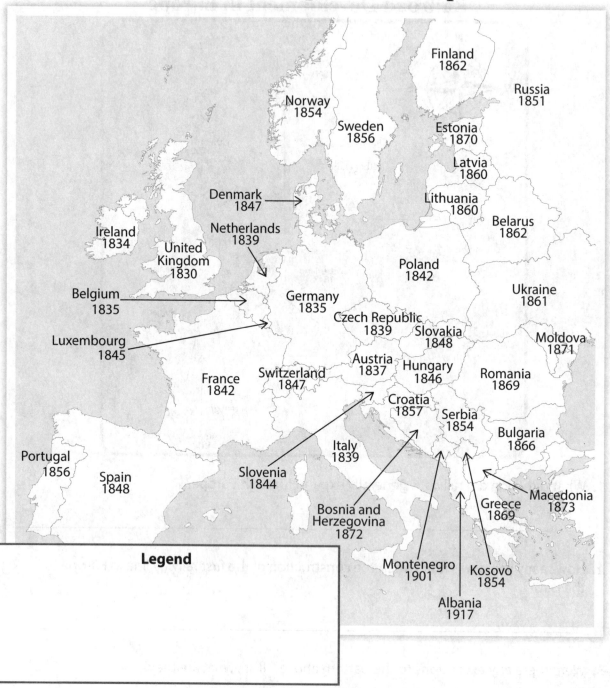

Finland
1862

Russia
1851

Norway
1854

Sweden
1856

Estonia
1870

Latvia
1860

Denmark
1847

Lithuania
1860

Belarus
1862

Ireland
1834

Netherlands
1839

United
Kingdom
1830

Poland
1842

Ukraine
1861

Belgium
1835

Germany
1835

Czech Republic
1839

Slovakia
1848

Moldova
1871

Luxembourg
1845

Austria
1837

Hungary
1846

Romania
1869

France
1842

Switzerland
1847

Croatia
1857

Serbia
1854

Bulgaria
1866

Italy
1839

Portugal
1856

Spain
1848

Slovenia
1844

Bosnia and
Herzegovina
1872

Greece
1869

Macedonia
1873

Montenegro
1901

Kosovo
1854

Albania
1917

Legend

1. Use colors to show which countries developed railway systems in each of the following periods: 1830s, 1840s, 1850s, 1860s, 1870s, and 1900–1920.

2. Add a legend to the map to show what your colors represent.

Name: _____ Date: _____

Directions: Read the text, and study the photo. Then, answer the questions.

Trains Connect Europe

The first steam locomotive was designed in the late 1700s. In the 1830s, railroads started expanding throughout Europe and the United States. In Europe, railways began in Britain. Trains were first used to transport industrial supplies. Soon after, trains carried passengers.

Not everyone liked the early railroads. Some people worried about the safety of trains. Others complained about the effects on the natural landscape. Trains were a major factor in the growth of factories and industry. Trains were particularly useful in Russia. Road travel was difficult in Russia because of the harsh winters. By the early 1900s, trains connected all of Europe.

Today, train travel is still a popular choice in Europe. Train passes are available for use across 28 countries in Europe. High-speed trains now travel up to 250 miles (400 km) per hour. It's even possible to travel underwater by train to get from Britain to France through a long tunnel called the Chunnel. With so much progress in less than two centuries, the next advance could be even faster.

1. Where did the first railways begin?

_____

2. What benefited most from railways?

_____

3. Why did people criticize the early railroads?

_____

_____

4. What is train travel in Europe like today?

_____

_____

_____

Name: _____ Date: _____

Think About It

Directions: Study the table, and answer the questions.

Railway Travel (in millions of kilometers traveled)		
Country	1980 (in millions)	2015 (in millions)
Austria (WE)	7,380	11,684
Belgium (WE)	6,963	10,333
Bulgaria (EE)	7,055	1,552
France (WE)	54,660	84,682
Ireland (WE)	1,032	1,916
Italy (WE)	39,587	39,290
Netherlands (WE)	8,910	17,770
Poland (EE)	46,300	7,486
Portugal (WE)	6,077	3,625
Romania (EE)	23,220	4,911
Spain (WE)	13,527	25,660

(WE) = Western Europe   (EE) = Eastern Europe

1. Which country had the most train travel in each year?

   1980: _____

   2025: _____

2. What patterns do you notice for each region (Western Europe and Eastern Europe)?

   _____

   _____

   _____

   _____

3. Do you think railroads will continue to expand in Europe? Why or why not?

   _____

   _____

   _____

   _____

Name: _____ **Date:** _____

Directions: Have you ever been on a train? If you could take a long train trip anywhere in the world, where would you go, and why? Write a journal entry about a real or imagined experience on a train.

_____

_____

_____

_____

_____

_____

_____

_____

_____

_____

_____

_____

_____

_____

_____

_____

_____

_____

_____

_____

_____

_____

Geography and Me

Reading Maps

Name: _____ Date: _____

Directions: This map shows the British Isles. Study the map, and answer the questions.

1. What four regions (also called *countries*) make up the nation of the United Kingdom?

   _____

   _____

2. Use cardinal directions to describe the location of the prime meridian on this map.

   _____

   _____

3. How far is France from the closest point in the British Isles?

   _____

4. How far is France from the farthest point in the British Isles?

   _____

Name: _____ **Date:** _____

Directions: Follow the steps to design a map that shows multiple types of information.

1. Color the United Kingdom (includes England, Scotland, Wales, and Northern Ireland).

2. Use a pattern to show Great Britain (includes England, Scotland, and Wales).

3. Use a different color or pattern to show the nation of Ireland (does not include Northern Ireland).

4. Create a legend to show what your patterns and colors mean.

Read About It

Name: _____ Date: _____

Directions: Read the text, and study the photo. Then, answer the questions.

Independence for Scotland?

Scotland and England have shared a government since the 1600s. But the Scots have not always been happy about it. In 2014, residents of Scotland had the chance to vote on this question: *Should Scotland be an independent country?* Supporters believed Scotland should be ruled by the Scots. They also believed an independent Scotland would be richer. People voting "no" believed Scotland was stronger as part of the United Kingdom.

On September 18, almost 85 percent of people in Scotland turned out to vote. About 55 percent voted no. Only 45 percent voted yes. People under 25 years old were more likely to vote yes. Those over 65 were much more likely to vote no. Scotland did not separate from the United Kingdom. But there are still groups within Scotland who are pushing for another vote. The future is yet to be determined.

1. What clues in the picture tell you which side the marchers were taking on independence?

   _____

   _____

   _____

2. What are some of the reasons for Scotland to leave the United Kingdom?

   _____

   _____

   _____

3. If Scotland had left the United Kingdom, how would the map of the British Isles change?

   _____

   _____

Name: _____ Date: _____

Directions: This graph shows how residents of Scotland viewed their national identity in 2011. Use the graph to answer the questions.

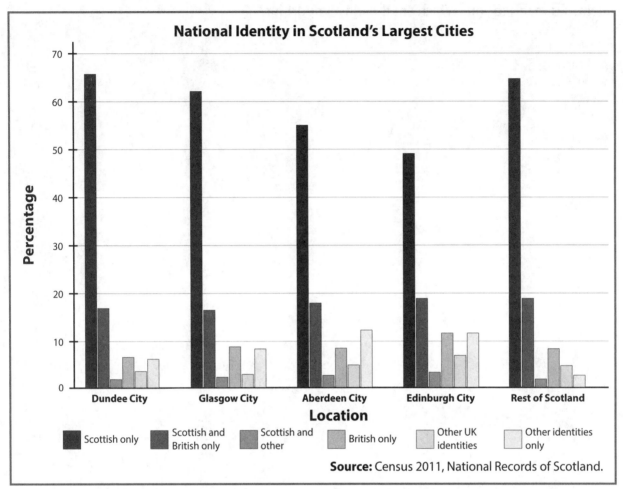

National Identity in Scotland's Largest Cities

Legend:
- Scottish only
- Scottish and British only
- Scottish and other
- British only
- Other UK identities
- Other identities only

Source: Census 2011, National Records of Scotland.

1. In Dundee City, what percentage of residents in the rest of Scotland viewed themselves as Scottish only? What percentage viewed themselves as British only?

   _____

2. What was more common in this census—a joint Scottish and British identity or a British-only identity?

   _____

3. Would you have expected the majority of people to vote for Scottish independence in 2014? Why?

   _____

   _____

Name: _____ Date: _____

Geography and Me

Directions: Think about the different kinds of geographic identities that you have: your neighborhood, community, region, and country. Draw or write to describe what those identities mean to you. Which are the most important?

_____

_____

_____

_____

_____

_____

_____

_____

Name: _____ Date: _____

Directions: This is a map of Mexico, Central America, and the Caribbean. Study the map, and answer the questions.

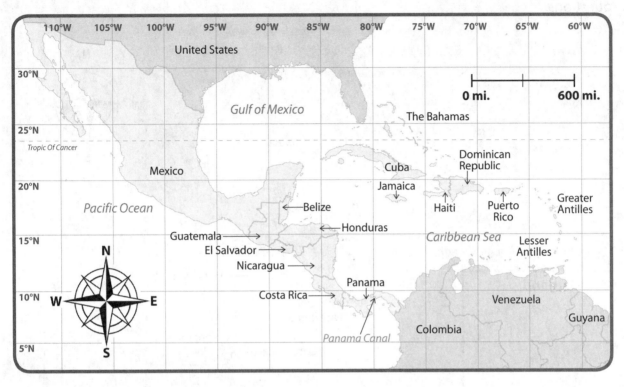

1. Use the scale to estimate the length of Mexico's border with the United States.

_____

2. All of Central America is located south of the Tropic of Cancer. What can you assume about the climate and vegetation of Central American countries?

_____

_____

3. What country lies at 20°N, 100°W?

_____

4. What country lies at 15°N, 85°W?

_____

Creating Maps

Name: _____   Date: _____

Directions: Use the chart as a guide to create a climate map of Central America. Use a different color for each type of climate. Be sure to include a legend. Then, answer the questions.

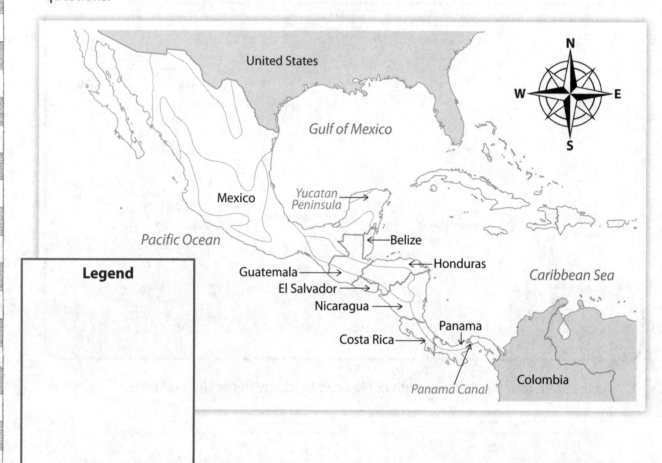

Type of Climate	Location
tropical wet	from Panama to Mexico along the Caribbean Coast
tropical wet and dry	from Panama to Mexico along the Pacific coast and the Yucatan Peninsula
highland	the central portion of Mexico and Central America
semiarid	north-central Mexico and the tip of Baja California
desert	Baja California and a small portion of northern Mexico

1. Which country has the most climate diversity?

_____

2. Which country has the least climate diversity?

_____

Name: _____ Date: _____

Directions: Read the text, and study the photo. Then, answer the questions.

A Disappearing Lake

The Aztecs built a powerful empire in central Mexico. They ruled in the 15th and 16th centuries. Around five to six million people lived in the Aztec empire. Their capital city was Tenochtitlán. It was built on islands in the middle of Lake Texcoco.

Lake Texcoco was a very large lake. It was connected to four other lakes in the valley. As the empire grew, the people needed more land. The Aztecs built more islands in the lake out of mud and dirt. These new artificial islands were called *chinampas*. The Aztecs used the land mostly for farming. After the Spaniards conquered the Aztecs in 1521, they drained the water from the lake to create more farmland. The soil turned out to be bad for farming, and the main result was the loss of all the water in the lake.

Today, Mexico City sits on the site of the ancient capital. The old lakes are now dusty and dry. The city is sinking a little each year. Drinking water and sewage treatment are huge problems for the modern city. The Mexican government is working to find a good solution. Maybe they will follow the example of the Aztecs and find a new innovation to solve the problem.

1. Why might the Aztecs have chosen to build their capital on Lake Texcoco?

_____

_____

2. What happened to the water in Lake Texcoco?

_____

_____

3. How did the actions of the Aztecs and the Spanish affect the modern residents of the area?

_____

_____

Think About It

Name: _____ **Date:** _____

Directions: This is a sixteenth century map of the Aztec capital city, Tenochtitlán. Study the map, and answer the questions.

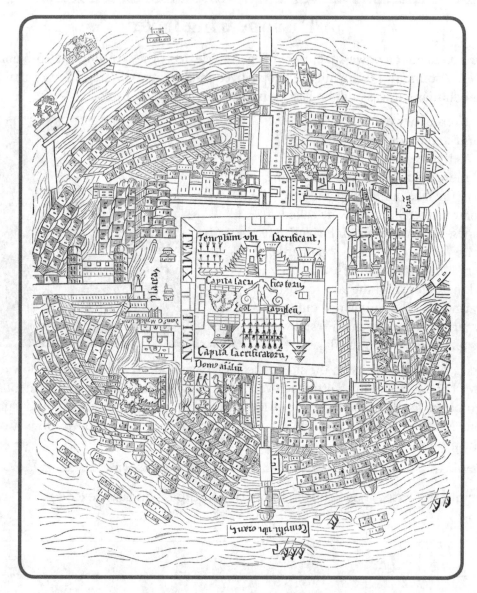

1. What notable geographic features are shown on this map?

_____

_____

2. What do you think the artist thought of the city of Tenochtitlán? Why?

_____

_____

Name: _____ **Date:** _____

Directions: There are only a few detailed accounts of Aztec culture and the city of Tenochtitlán. Draw a picture of how you imagine life in the city before the Europeans arrived.

Geography and Me

Reading Maps

Name: _____   Date: _____

Directions: Study the map of Canada, and answer the questions.

1. What latitude line is the northern border for four different Canadian provinces? List those Canadian provinces.

_____

_____

_____

2. Where is the majority of the Canadian population located? How do you know?

_____

_____

_____

3. At approximately what latitude is most of the shared border with the United States?

_____

Name: _____ Date: _____

Directions: Canada has ample freshwater from lakes and rivers. But, the rivers and lakes are not evenly distributed across the country. Follow the steps to show the distribution of this important resource.

Legend

1. Outline the major rivers and lakes in Canada.

2. Use the same color in different shades to show which provinces and territories have the most to least freshwater.

3. Create a legend to show what your shades and outlines represent.

Read About It

Name: _____ Date: _____

Directions: Read the text, and study the photo. Then, answer the questions.

A Great Lake

Canada is home to water in many forms. There are approximately 77,000 square miles (124,000 square km) of ice and glaciers. There are many rivers and large lakes, including the five Great Lakes, shared with the United States. Finally, Canada is surrounded on three sides by oceans and bays.

One of these lakes is the eighth largest in the world. Great Bear Lake is located in the Northwest Territories. It covers more than 12,000 square miles (19,000 square km), part of which is inside the Arctic Circle. Great Bear Lake was named a UNESCO Biosphere Reserve. This helps with both conservation of the lake and the economic growth of Deline, an aboriginal town next to the lake. The idea is to balance the people's needs with the lake's needs.

aerial view of Great Bear Lake

To the people of the Sahtuto'ine culture, the lake is more than a pretty view. They believe it is a living thing and an essential part of their culture. A Sahtuto'ine elder named Eht'se Ayah even proclaimed a prophecy about Great Bear Lake. He said that near the end of the world, people would come to Great Bear Lake because there was no other food or water on Earth. Under this view, conservation efforts at Great Bear Lake are a necessity for all of us. That just might be true either way.

1. What makes Great Bear Lake unique?

_____

_____

2. What did a Sahtuto'ine elder Eht'se Ayah proclaim about Great Bear Lake?

_____

_____

3. What do you think is the most important form of freshwater, and why?

_____

_____

Name: _____ Date: _____

Directions: The table shows available freshwater and the amount used each year. Study the table, and answer the questions.

	Available Freshwater		Water Used	
	Per year (cubic kilometers)	Per person (cubic meters)	Per year (cubic kilometers)	Per person (cubic meters)
Australia	492	20,527	20	824
Brazil	8,647	41,603	75	370
Canada	3,478	103,899	38	1,078
China	2,840	2,018	554	411
Egypt	58	637	78	911
Germany	154	1,909	33	411
India	1,911	1,458	761	602
Russia	4,525	31,543	66	456
United States	3,069	9,538	486	1,543

Source: Statistics Canada

1. How does Canada compare to the other countries on the list in terms of how much water it has available?

_____

_____

2. Why might Canadians use so much water?

_____

_____

_____

3. Compare how much water each country uses to the water it has available. Which country is in a water crisis, and why?

_____

_____

_____

Name: _____ **Date:** _____

Directions: Imagine that you lost your current source of water. Brainstorm three options you might have for getting drinking water. What are the pros and cons of each option? Record your ideas in the table.

Option	Pros	Cons

Geography and Me

Name: _____ **Date:** _____

Directions: Study the map, and answer the questions.

Monarch Migrations

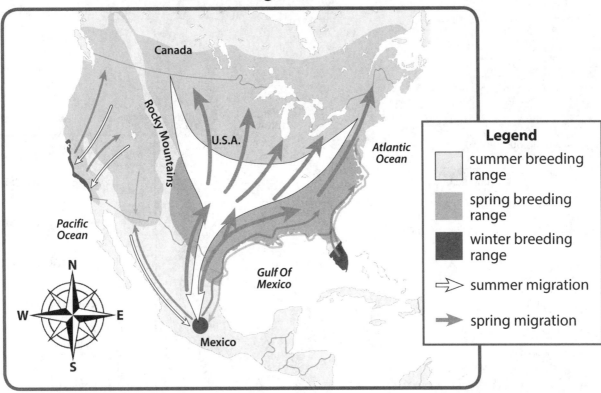

1. Find the areas where monarch butterflies spend their winters. Why do you think they spend winters in those areas?

   _____

   _____

2. What can you assume is important for monarch butterflies based on their migration patterns and the seasons? What geographic evidence can you use?

   _____

   _____

   _____

3. What physical feature do monarch butterflies seem to avoid? Why might they avoid it?

   _____

   _____

Name: _____ Date: _____

Directions: Follow the steps to complete the map.

North American Climates

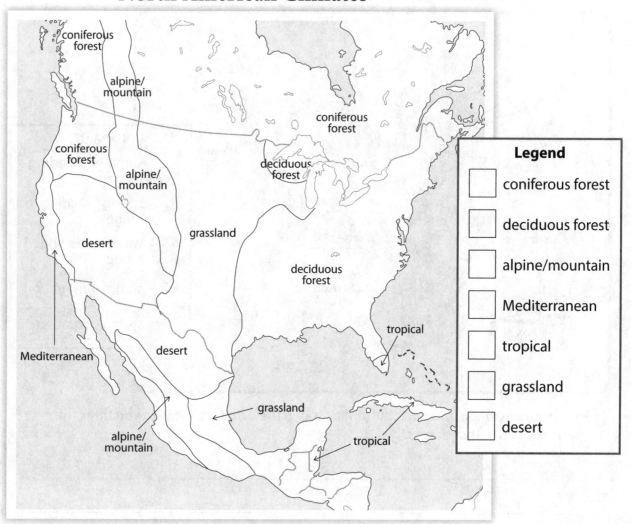

Legend

☐ coniferous forest

☐ deciduous forest

☐ alpine/mountain

☐ Mediterranean

☐ tropical

☐ grassland

☐ desert

1. Color each type of climate a different color on the map.

2. Color the legend to match the colors you used on the map.

3. Describe the locations of the following climates: Mediterranean, desert, and tropical.

_____

_____

_____

_____

Name: _____ Date: _____

Directions: Read the text, and study the photo. Then, answer the questions.

Endangered Butterflies

Monarch butterflies used to be easy to find. Today, they are in danger of becoming extinct. The orange, black, and white markings of the monarch butterfly are unique. These butterflies migrate, so they are common to most parts of the United States and Canada. However, in recent years, the monarch's habitat has been hit hard by human and natural destruction. Milkweed is their favorite food, and it is where they lay their eggs. But milkweed has also become harder to find. The monarch is now in trouble.

The problem is growing every year. Scientists estimate that the number of monarchs who completed their annual migration in 2013 was just six percent of the number from 10 years earlier. That's a huge loss—not only for the monarchs. Most plants need insects to pollinate them. Animals depend on the health of plants. Our environment needs pollinators like monarchs to stay healthy and in balance.

There are actions people can take to help the monarchs. One way is to plant milkweed along the roadside. Another is to restrict the use of herbicides that kill milkweed. But perhaps the easiest and most important way to help is to plant milkweed in our own backyards to encourage monarchs to visit.

1. What are two reasons for the decline in monarch butterflies?

_____

_____

2. Why is milkweed so important to monarch butterflies?

_____

_____

3. What is the best way people can help save the monarch butterflies?

_____

_____

Think About It

Name: _____ Date: _____

Directions: This bar graph shows the estimated population of monarch butterflies from 1995 to 2016. Study the graph, and answer the questions.

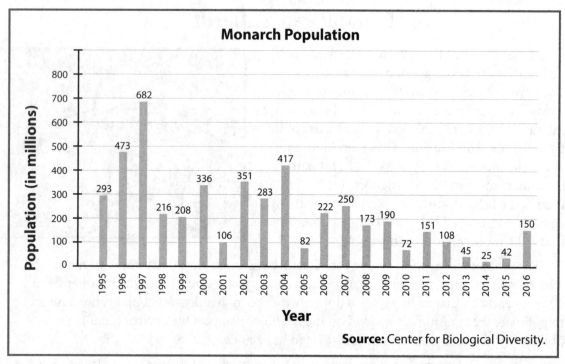

Source: Center for Biological Diversity.

1. What patterns do you see in monarch populations?

_____

_____

2. When was the sharpest decrease in population? What do you think accounts for that decrease?

_____

_____

3. Given this data, what do you predict will happen over the next few years? What does it depend on?

_____

_____

_____

Name: _____ **Date:** _____

Directions: Write a letter to the editor of a local newspaper about the endangered monarch butterflies. Be sure to explain the problem and what you think people should do about it.

_____

_____

_____

_____

_____

_____

_____

_____

_____

_____

_____

_____

_____

_____

_____

_____

_____

_____

_____

_____

_____

Geography and Me

Reading Maps

Name: _____ **Date:** _____

Directions: This is a political map of the Caribbean. Study the map, and answer the questions.

The Bahamas
Bimini
Islands

North Atlantic Ocean

Ragged Island
Range
Cuba

Turks and
Caicos Islands
(U.K.)

British Virgin Islands
(U.K.)

Virgin Islands
(U.K.)

Cayman Islands
(U.K.)

Navassa Island
(U.S.)

Haiti

Dominican
Republic

Puerto Rico
(U.S.)

Anguilla (U.K.)
Saint-Martin (France)
Saint Barthelemy (France)

Jamaica

Saba and Sint Eustatius (Neth.)
Saint Kitts and Nevis
Montserrat (U.K.)

Antigua and Barbuda

Guadeloupe (France)

Dominica

Caribbean Sea

Isla De Avés
(Venezuela)

Martinique (France)

Saint Lucia

Isla De La Providencia
(Colombia)

Curacao
(Neth.)

Aruba
(Neth.)

Bonaire
(Neth.)

Saint Vincent and
the Grenadines

Barbados

Islas
Del Maiz

Isla De San Andrés
(Colombia)

Grenada

Trinidad and
Tobago

1. What do these countries have in common?

_____

2. What do you notice about many of the smaller islands, such as Martinique and St. Martin?

_____

_____

_____

3. What are the five largest island countries in the Caribbean?

_____

_____

_____

Name: _____ **Date:** _____

Directions: Follow the steps to complete the map.

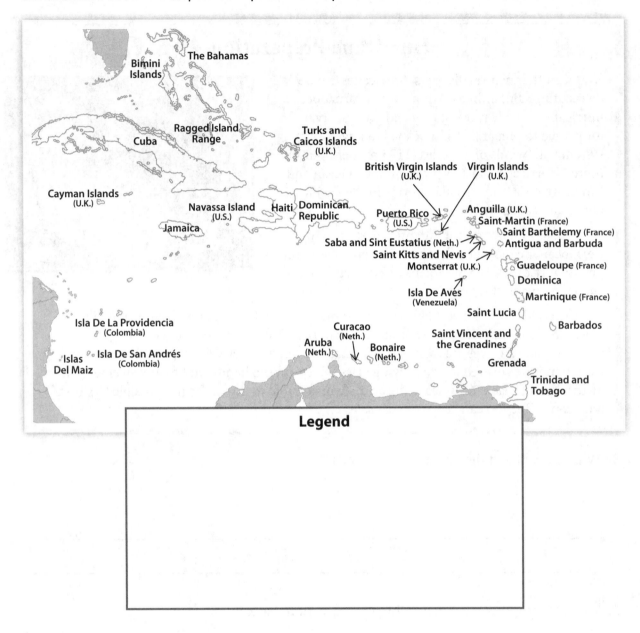

1. Use different colors to shade each of the following groups: United States territories, United Kingdom territories, French territories, and Dutch territories.

2. Create a legend to show what the colors represent.

3. Add a compass rose to the map.

4. Label the bodies of water.

Read About It

Name: _____ Date: _____

Directions: Read the text, and study the photo. Then, answer the questions.

Hurricane Preparation

Like other natural disasters, hurricanes can be devastating. But unlike earthquakes or tornadoes, hurricanes usually move slowly and give plenty of time for people to prepare. A hurricane is a large storm system with winds of 74 miles (119 km) per hour or more. Hurricanes start over the Atlantic Ocean and can move over land. Similar storms in the Pacific Ocean are called *typhoons*.

People who live on the Atlantic coast or in the Caribbean are most at risk for injuries and loss of life due to hurricanes. These storms can damage property and cause major flooding.

The good news is that there are ways to prepare. First, you should have a general emergency kit, including water, food, and batteries. Make sure your house is in good condition and all trees are trimmed. Get a generator, or otherwise prepare for power loss. Keep windows covered to avoid flying glass. Keep a radio handy, and follow all emergency instructions. Evacuate, or leave the area, if you are told to do so. It's not possible to stop a hurricane. But you can keep yourself safe.

1. What are some of the effects of hurricanes?

_____

_____

_____

2. Why are the Caribbean Islands threatened by hurricanes?

_____

_____

3. How can people prepare for a hurricane?

_____

_____

Name: _____ Date: _____

Directions: These pictures show damage following a hurricane. Study the pictures, and answer the questions.

1. What damage do you see in the photos?

   _____

   _____

2. How could this kind of damage affect people?

   _____

   _____

   _____

3. Which effect of a hurricane do you think would be most difficult to protect against, and why?

   _____

   _____

   _____

4. What do these photos tell you about the power of hurricanes?

   _____

   _____

   _____

Name: _____ **Date:** _____

Directions: What natural disasters are common in your region? What do you do to prepare for them? What additional preparation could you do? Complete the chart to answer these questions.

Natural Disaster	Preparation

Geography and Me

Name: _____ Date: _____

Directions: This is a physical map of Mexico and Central America. Study the map, and answer the questions.

1. What are the two peninsulas in Mexico?

_____

2. What body or bodies of water surround each peninsula?

_____

_____

3. What is the largest lake in Central America? In which country is it located?

_____

_____

4. Describe the locations of major mountain ranges.

_____

_____

Creating Maps

Name: _____ **Date:** _____

Directions: Use symbols to show the top two exports in Central American countries. Be sure to include a legend.

Legend

Country	Top Exports in 2015
Guatemala	bananas, sugar
Honduras	coffee, textiles
El Salvador	textiles, electrical capacitors
Belize	sugar, bananas
Nicaragua	insulated wire, textiles
Panama	ships, coal, tar
Costa Rica	medical instruments, bananas

Name: _____ **Date:** _____

Directions: Read the text, and study the photo. Then, answer the questions.

Fair Trade in Central America

Central America exports coffee, fruit, and sugar to other countries. These are cash crops. Cash crops are grown to trade and make money. Countries that cannot easily grow these crops import them from Central American countries.

Many foreign companies moved into the region to do business. For example, the United Fruit Company owned banana plantations in many countries. Foreign companies ended up with control over the production and the profits. In addition, locals have criticized these companies for damaging the land, exploiting workers, and interfering in local politics.

In recent years, companies and governments have tried to solve some of those problems. A movement called Fair Trade began. Fair Trade companies buy from groups of farmers. These farmers are guaranteed enough money to cover their costs and still have money to live on. Farmers have to treat the land and their workers well. The goal is for everyone to benefit fairly.

It is still too early to see if the Fair Trade Movement will improve the lives of farmers. But it is important to find a way to do business that benefits both the people and the land.

1. What are the main cash crops in Central America?

_____

_____

2. Why were companies like United Fruit criticized?

_____

_____

3. Do you believe the Fair Trade movement will work? Explain your answer.

_____

_____

Name: _____ Date: _____

Think About It

Directions: This table shows the value of all exports from several Central American countries to the United States. Study the table, and answer the questions.

Annual Exports to the United States (in dollar value)				
Country	1985	1995	2005	2015
Guatemala	409,000,000	1,526,800,000	3,137,382,263	4,120,741,294
El Salvador	395,600,000	812,200,000	1,988,765,718	2,531,730,305
Honduras	375,400,000	1,441,200,000	3,749,243,099	4,759,684,807
Costa Rica	501,300,000	1,843,200,000	3,415,279,262	4,488,909,896
Panama	410,600,000	307,100,000	327,060,069	408,339,232

1. What happened to exports from these countries between 1985 and 2015?

_____

_____

2. Which country had the greatest increase of exports to the United States from 1985 to 2015?

_____

3. What are some possible reasons for such a large increase over the years?

_____

_____

_____

_____

4. What do you predict will happen to the amount of exports to the United States in 2025? Use the chart to support your answer.

_____

_____

_____

_____

Name: _____ **Date:** _____

Directions: Think about the products exported from Central America. How would your life change if these products became more expensive? What if they became less available as well? Use the table to record your thoughts.

Item	Effects of Price Increase	Effects of Less Availability
bananas		
sugar		
textiles		
medical instruments		

© Shell Education

Reading Maps

Name: _____ **Date:** _____

Directions: Use the map of South America and the scale to answer the questions.

1. What is the largest country in South America?

_____

2. Name a mountain range in South America.

_____

3. What is the approximate distance from the northernmost point of South America to the southernmost point?

_____

Name: _____ Date: _____

Directions: Use symbols or shading on the map to show the main language spoken in each South American country. Add your symbols or shading to the legend.

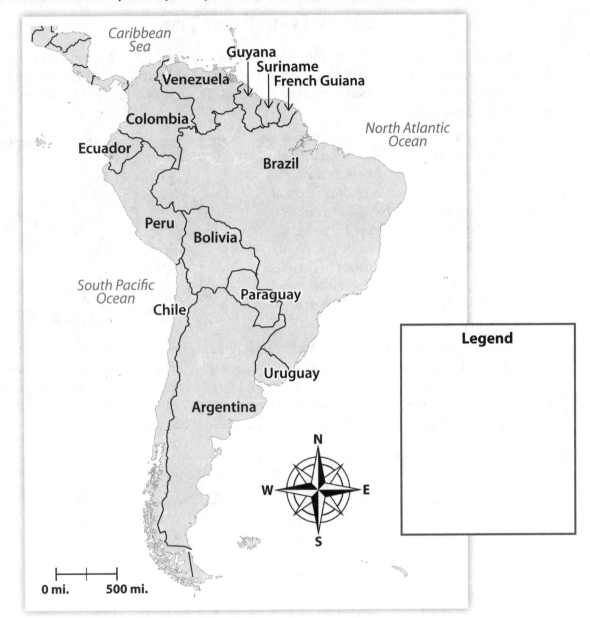

Country	Language
Argentina	Spanish
Bolivia	Spanish
Brazil	Portuguese
Chile	Spanish
Colombia	Spanish
Ecuador	Spanish
French Guiana	French

Country	Language
Guyana	English
Paraguay	Spanish
Peru	Spanish
Suriname	Dutch
Uruguay	Spanish
Venezuela	Spanish

Read About It

Name: _____ Date: _____

Directions: Read the text, and study the photo. Then, answer the questions.

A Very Long Road Trip

It is possible to drive all the way from Alaska to the southern tip of Argentina. In 1923, the idea for the Pan-American Highway was introduced. The highway would connect all of North and South America with one single roadway. That's 30,000 miles (48,000 km) of road—the longest in the world.

The Pan-American Highway has taken a long time and a lot of money to build. The United States helped fund some of the roads through Central America. The road is really a series of roads with a few detours. To pass the Straits of Magellan, you must take a ferry. There are also bridges and tunnels involved at various points. The trickiest part is a tropical forest in between Panama and Colombia called Darien National Park. It is possible to drive through the park, but it is slow and dangerous.

The road is not complete. It is a continuous work in progress. Supporters try to get money to finish the road. However, the remaining area is wild jungle and forest. A paved road would destroy part of the preserved land. To most local people, it has not seemed worthwhile. So, for now, travelers will have to use a ferry to continue down the highway.

1. What is the Pan-American Highway?

_____

_____

2. Why is the Pan-American Highway incomplete?

_____

_____

3. What are the advantages of the Pan-American Highway?

_____

_____

_____

Name: _____ Date: _____

Directions: South America is a large continent. Study the photos and chart. Then, answer the questions.

Continent	Land (sq. km)	Number of Countries
Asia	17,212,000	44
Africa	11,668,599	54
North America	9,540,000	23
South America	6,890,000	12
Antarctica	5,400,000	0
Europe	3,930,000	47
Australia/Oceania	3,291,903	14

1. How does South America compare to the rest of the world in terms of overall land mass?

_____

_____

2. What does the total number of countries in South America compared to the other continents show?

_____

_____

3. How do the photos support the idea that South America is a large continent?

_____

_____

Name: _____ **Date:** _____

Geography and Me

Directions: Draw a map of your country. Label your community. Then, label five places you have been or would like to visit.

Name: _____ Date: _____

Directions: This is a map of South America with the Amazon River Basin shaded. Study the map, and answer the questions.

Amazon River Basin

1. Through which countries does the Amazon River run?

_____

_____

2. Which countries are included in the Amazon River Basin?

_____

_____

3. Using the scale, approximately how long is the Amazon River?

_____

Creating Maps

Name: _____ Date: _____

Directions: To understand the size of the Amazon rainforest, compare it to the size of the United States. Using the measurements, draw an approximate outline of the continental United States in the box. Use the same scale as the map of South America.

Amazon River Basin

Caribbean Sea
Venezuela
Guyana
Suriname
French Guiana
Colombia
Orinoco River
North Atlantic Ocean
Ecuador
Negro River
Amazon River
Madeira River
Marañon River
Brazil
Xingu River
Araguaia River
Tocantins River
Peru
Bolivia
South Pacific Ocean
Chile
Paraguay
Argentina
Uruguay

0 mi.   500 mi.

Measurements of the United States
2,680 miles wide (east to west)
1,582 miles long (north to south)

Name: _____   Date: _____

Directions: Read the text, and study the photo. Then, answer the questions.

Amazon in Peril

Deforestation is the removal and destruction of part of a forest. Environmentalists have succeeded in calling attention to deforestation and endangered species in the Amazon. For years, their efforts have helped decrease deforestation.

The Amazon Rainforest is the largest rainforest in the world. It is home to millions of species. There are jaguars, sloths, and pink dolphins. Many of these species live only in the Amazon. Their habitat is in danger. There are also about 390 billion trees in the rainforest. These trees hold huge amounts of carbon. This helps keep harmful greenhouse gases out of the air. But that carbon is released into the air when the trees die.

The greatest cause of deforestation in the Amazon is cattle ranching and large-scale farming. People burn huge swaths of forest every year to make room for farms and ranches. This contributes to air pollution, and it releases greenhouse gases.

Without a healthy Amazon, it's not just plants and animals that are in danger. Our people and our planet will be as well.

1. What makes the Amazon unique?

_____

2. What are some of the reasons for deforestation in the Amazon?

_____

_____

3. What could happen if deforestation is allowed to continue?

_____

_____

Name: _____ Date: _____

Think About It

Directions: This table shows how much of the Amazon rainforest has been cleared since 1970. Study the table, and answer the questions.

Period	Remaining Rainforest (sq. km.)	Annual Forest Loss (sq. km.)	Percentage of Rainforest Left
pre-1970	4,100,000	n/a	n/a
1990	3,692,020	13,730	90.0%
2000	3,524,097	18,226	86.0%
2010	3,358,788	7,000	81.9%
2014	3,336,896	5,012	81.4%
2015	3,330,689	6,207	81.2%
2016	3,322,700	7,989	81.0%

1. Approximately how much of the forest cover in the Brazilian Amazon was lost from 1970 to 2010?

_____

2. Looking at the annual forest loss, what might have happened between 2000 and 2010?

_____

_____

3. Based on the last few years in the table, what do you predict will happen to the rainforest in the near future?

_____

_____

4. What next steps do you think leaders should take to protect the Amazon Rainforest?

_____

_____

5. What can you do to help protect the Amazon Rainforest?

_____

_____

Name: _____ **Date:** _____

Directions: Summarize what you learned about the Amazon this week. What did you find most interesting, and why?

_____

_____

_____

_____

_____

_____

_____

_____

_____

_____

_____

_____

_____

_____

_____

_____

_____

_____

_____

_____

_____

_____

Geography and Me

Reading Maps

Name: _____ **Date:** _____

Directions: This is a physical map of Peru. The darker colors show higher elevations. Study the map, and answer the questions.

1. What mountain range runs through Peru?

   _____

2. Use cardinal directions to describe where the flattest part of Peru is located.

   _____

   _____

3. Name the major rivers of Peru.

   _____

   _____

   _____

Name: _____ **Date:** _____

Directions: Follow the steps to complete the map.

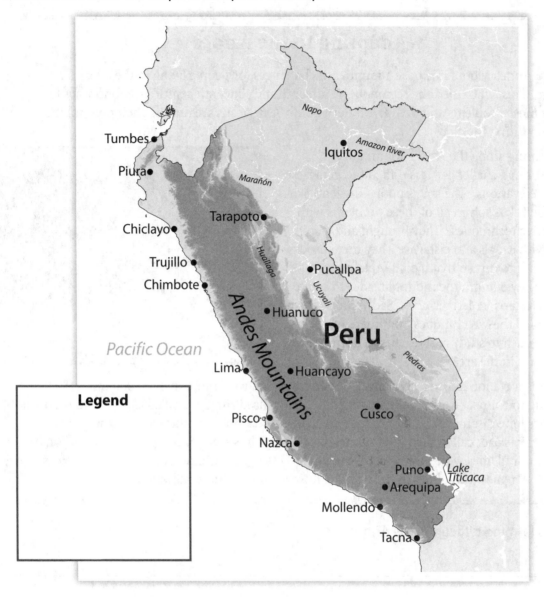

1. Draw an outline around the Andes Mountains.

2. The portion of Peru east of the Andes Mountains is part of the Amazon River basin. Color this area.

3. Use a symbol to identify Cusco as the capital city of the ancient Inka empire.

4. Use a symbol to identify Lima as the modern-day capital.

5. Create a legend to show what your colors and symbols represent.

Read About It

Name: _____ Date: _____

Directions: Read the text, and study the photo. Then, answer the questions.

Adapting to the Andes

The Andes Mountains pose a number of challenges for people living in the area. The ancient Inkas, who ruled the region in the fifteenth and sixteenth centuries, found ways to handle these problems and thrive. Modern South Americans are finding their own solutions, while revisiting the old.

Farming along the Andes Mountains has always been a challenge. It is difficult to get water to high elevations. The soil is poor. The land is not flat. The Inkas solved all of these problems with innovative techniques. They brought and stored water with canals and cisterns. They mixed rocks and sand into dirt so that it could retain more water. They cut into the mountainsides to create huge, flat steps, called terraces. Sadly, the Spanish abandoned these techniques. But there has been a recent push to study and use Inkan farming techniques in the area.

Transportation is also a challenge. The Inkas built a road more than 20,000 miles (32,000 km) long to connect their mountain kingdom. They used runners, called *chasquis*, to transmit messages long distances and llamas to carry loads. Today in Peru, there is a highway close to the old Inka road, called the Longitudinal de la Sierra. It is still not completely paved. Some significant rail lines were built when it became necessary for mining purposes. Peru boasts two of the highest train lines in the world. However, not all these railways connect.

1. Why is farming difficult in the Andes?

_____

_____

2. How did the Inkas solve the problems of farming in the Andes?

_____

_____

3. How has transportation in the Andes changed over time?

_____

_____

Name: _____ Date: _____

Directions: This graph shows how much land from each country is in the Andes Mountains. It also shows the percentage of the population that lives in the Andes. Study the graph, and answer the questions.

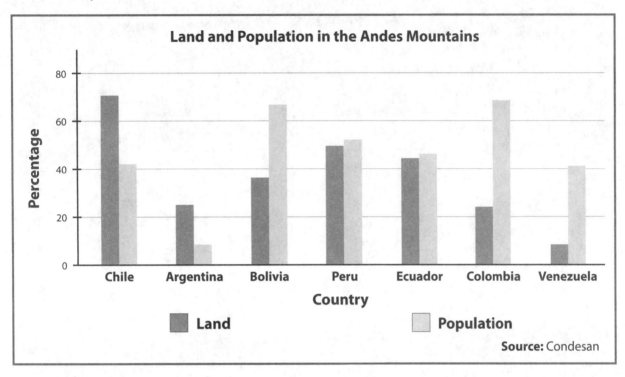

1. Which three countries have the greatest percentage of their land in the Andes?

   _____

   _____

2. Which three countries have the greatest percentage of their population in the Andes?

   _____

   _____

3. Which countries would you expect to have the most cities in the Andes, and why?

   _____

   _____

   _____

Think About It

Geography and Me

Name: _____ Date: _____

Directions: In the Inka Empire, there were no modern communication methods and no vehicles. They also did not use a writing system or the wheel. Imagine you lived in the Inka Empire. Draw or write three ways to transport messages and cargo.

Name: _____   Date: _____

Directions: Study the map, and answer the questions.

Top Exports

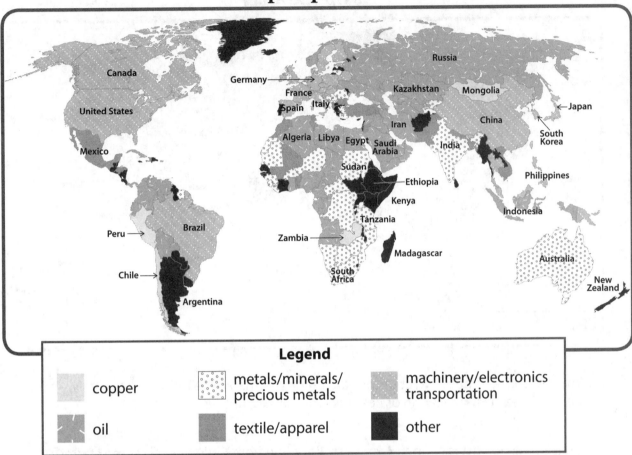

Legend

copper	metals/minerals/precious metals	machinery/electronics transportation
oil	textile/apparel	other

1. What type of products do China and the United States both export?

_____

2. What is the most common type of export worldwide? Where is it most common?

_____

_____

3. Which two South American countries are major copper producers?

_____

_____

Name: _____ **Date:** _____

Directions: Follow the steps to complete the map.

Top Exports

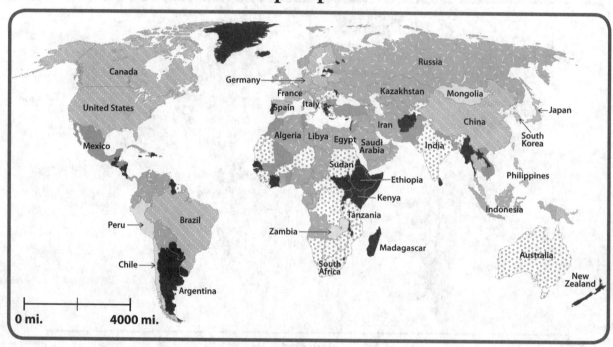

1. The top copper-importing countries are China, Germany, the United States, Italy, and South Korea. Draw a star on or near each of these countries.

2. Chile's main export is copper. Draw a straight line from Chile to each of the starred countries.

3. Use the scale to estimate the distance from Chile to each importing country.

_____

_____

_____

4. What does this tell you about Chile's trade network?

_____

_____

_____

Name: _____ **Date:** _____

Directions: Read the text, and study the photo. Then, answer the questions.

Copper in Chile

Chile is a long and narrow country. It runs along much of the western coast of South America. Chile has mountains, vineyards, and 90 active volcanoes. It is also rich in minerals. Chile produces more copper than any other country. In 2015, about half of Chile's exports were copper products. Copper has been in high demand in recent years. It is used in many consumer products.

Six of the world's largest copper mines are in Chile. One of them is the Chuquicamata copper mine. In 2010, Chuquicamata began building a new underground mine. This expansion will cost billions of dollars. A second mine, the Escondida mine, is even newer and larger. These mines bring money and jobs.

Chile has recently started to focus more on the effects of mining. Copper mining uses a lot of water. Chilean mines are often located in areas with limited water, such as the Atacama Desert. Deforestation and pollution are problems as well. All the people moved away from a town near Chuquicamata because of air pollution. However, Chile needs the money created from copper mining. The world needs copper for many products. The goal will be to balance the environment and the economy.

1. Why is copper important to Chile?

_____

_____

2. What evidence shows that Chile is a big copper producer?

_____

_____

3. What environmental problems does copper mining create?

_____

_____

Think About It

Name: _____ Date: _____

Directions: This graph shows copper prices from 2002 to 2014. Study the graph, and answer the questions.

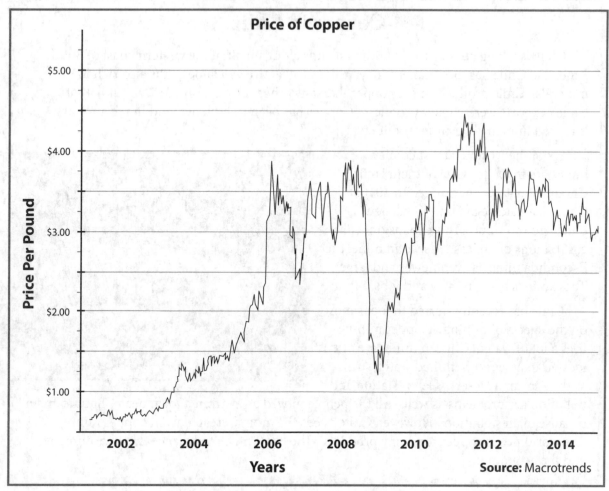

1. How stable is the price of copper from year to year?

   _____

   _____

2. What might happen to the Chilean economy when the prices for copper go up and down?

   _____

   _____

3. Should Chile expand its economy away from copper mining? Why or why not?

   _____

   _____

Name: _____ Date: _____

Directions: Copper is used in many products, including heaters, air conditioners, water pipes, copper wiring, computer chips, telephone wiring, and cars. Draw all the different ways you use copper.

Name: _____ Date: _____

Directions: This is a map of the Galápagos Islands. This is an archipelago, or a group of islands, in the Pacific Ocean that are part of Ecuador.

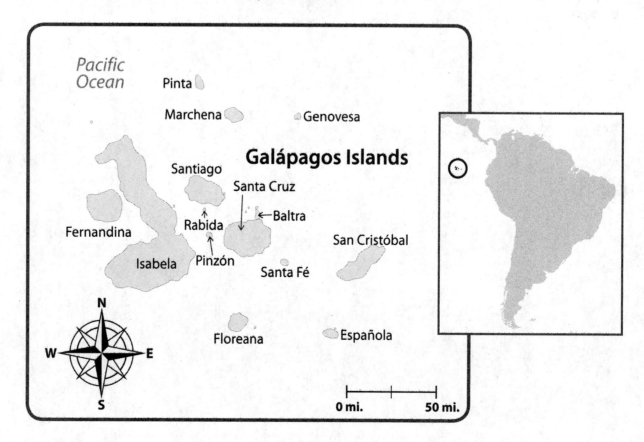

1. In which body of water are the islands located?

_____

2. What is the largest island of the Galápagos Islands?

_____

3. What is the smallest named island on this map?

_____

4. What do the names of the Galápagos Islands tell you about their history?

_____

_____

_____

Name: _____ **Date:** _____

Directions: The Galápagos Islands are most famous for the unique wildlife that inspired Charles Darwin's theory of evolution. Draw symbols on the map to show where different animals can be found. Be sure to include a legend.

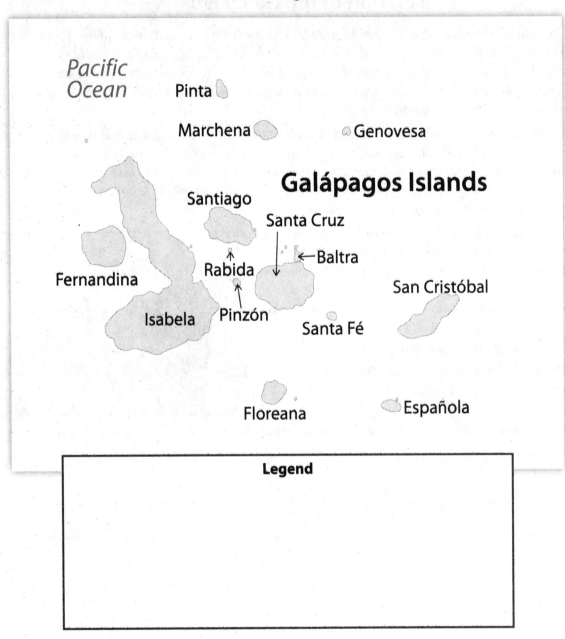

Legend

Giant Tortoise: San Cristóbal, Santa Cruz, and Isabela

Marine Iguana: Fernandina, Santiago, Isabela, and Santa Cruz

Galápagos Penguin: Isabela and Floreana

Name: _____ Date: _____

Read About It

Directions: Read the text, and study the photo. Then, answer the questions.

Isolation in the Galápagos

The Galápagos Islands are located hundreds of miles from the coast of Ecuador. There is no other land close to them. These islands were untouched by humans for most of history. Isolation made the Galápagos the perfect place for Charles Darwin's studies on evolution. Evolution is the theory that all life on Earth has slowly changed to adapt to its environment. Those who change, survive.

On each island, animals have changed according to their habitats. Each island has different lands and food sources. For example, the beaks on finches are different on each island. The shape of their beaks depends on what the finches eat. Iguanas also have different features, depending on their food sources. Iguanas that eat algae have different mouths and claws than iguanas that eat cactus.

The Galápagos Islands are now full of people. More than 200,000 tourists visit the islands every year. There are also more than 20,000 permanent residents. Settlers brought with them new plants and animals, such as donkeys, goats, and pigs. These new animals destroyed much of the native habitat.

Today, the Galápagos Islands are trying to find a balance. Tourism, study, and conservation are all important. The islands can no longer hide. People must protect them instead.

1. Why were the Galápagos isolated?

   _____

   _____

2. Why are the animals different from island to island?

   _____

   _____

3. What challenges do conservationists face today in protecting the land and animals of the Galápagos?

   _____

   _____

Name: _____ Date: _____

Directions: Study the photo, and answer the questions.

1. What does this picture tell you about the impact of tourists and settlers on the islands?

_____

_____

_____

2. Why do islands, especially the Galápagos, have an extra challenge in dealing with garbage?

_____

_____

_____

3. How could unattended garbage cause problems for the wildlife on the Galápagos?

_____

_____

_____

_____

Geography and Me

Name: _____ Date: _____

Directions: What wildlife is common in your area? Write and draw to describe four animals in your area. Wildlife can be small creatures, such as birds and insects, or they can be large, such as bears and deer, or anything in between.

28627—180 Days of Geography

Name: _____ Date: _____

Directions: This map shows some of the top oil-producing countries in 2016. The top three are labeled as 1, 2, and 3. Study the map, and answer the questions.

Top Oil Producers

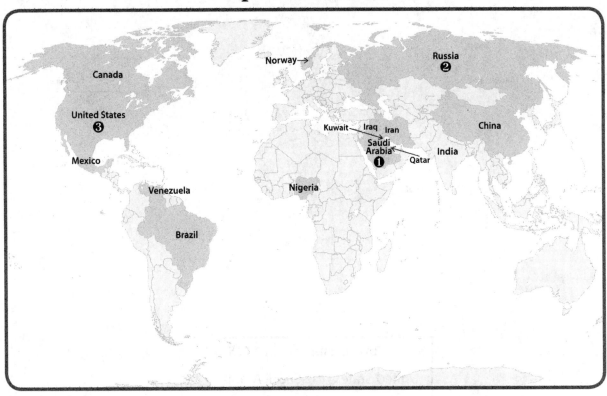

1. Which countries are the top producers of oil?

   _____

   _____

2. Which continent has the highest number of top oil-producing countries?

   _____

3. What does this map say about the role of oil in the world economy?

   _____

   _____

   _____

   _____

Creating Maps

Name: _____ **Date:** _____

Directions: Follow the steps to complete the map.

Top Oil Consumers

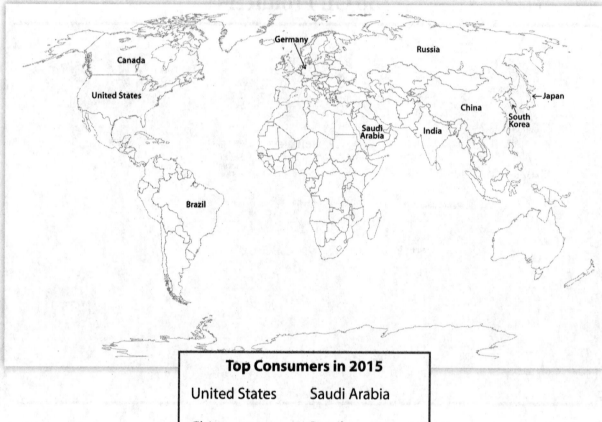

Top Consumers in 2015	
United States	Saudi Arabia
China	Brazil
India	South Korea
Japan	Canada
Russia	Germany

1. Use colors or shading to mark the top consumers of oil.

2. Label as many other countries as you can on the map.

3. Look up the names of five countries you did not know. Label them on the map in a different color.

© *Shell Education*

Name: _____ **Date:** _____

Directions: Read the text, and study the photo. Then, answer the questions.

Fossil Fuels

Fossil fuels come from the remains of prehistoric plants and animals. These fossils were buried and decayed over millions of years. Eventually, they turned into coal, oil, and natural gas. These fuels are used to power motors, run factories, heat houses, and create electricity. The great majority of power in the United States has come from fossil fuels for over a century. Globally, over 80 percent of all fuel sources were fossil fuels in 2015.

Fossil fuels are plentiful, and they are relatively cheap. We already have well-developed systems of extracting and distributing them. However, coal, oil, and natural gas have serious drawbacks. They are not renewable. This means that we cannot create more of them. Once we run out of fossil fuels, they are gone forever. They are also not clean. Fossil fuels pollute the air. They release significant amounts of carbon dioxide. More carbon dioxide in our atmosphere leads to warmer temperatures. So, many scientists say that we should start using alternative fuel sources long before we run out of fossil fuels.

There is no easy answer to the energy question. Changing our habits is hard and can be expensive. However, sooner or later, this is a problem that we must face.

1. What are fossil fuels, and where did they come from?

_____

_____

2. What are some advantages of fossil fuels?

_____

_____

3. What are some of the problems with fossil fuels?

_____

_____

Think About It

Name: _____ Date: _____

Directions: This graph shows how much petroleum (oil) is used to make different kinds of products in the United States. Study the graph, and answer the questions.

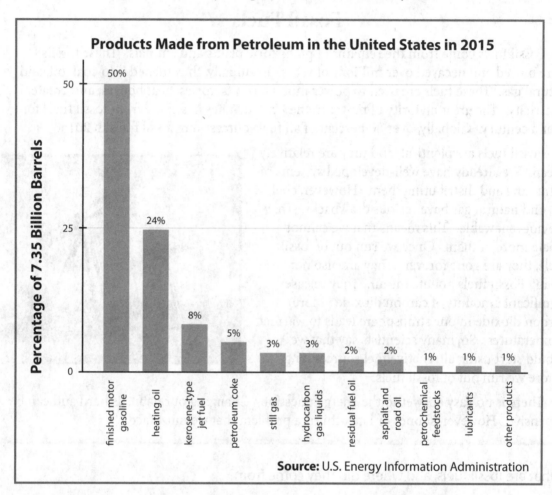

Products Made from Petroleum in the United States in 2015

Percentage of 7.35 Billion Barrels

50% finished motor gasoline
24% heating oil
8% kerosene-type jet fuel
5% petroleum coke
3% still gas
3% hydrocarbon gas liquids
2% residual fuel oil
2% asphalt and road oil
1% petrochemical feedstocks
1% lubricants
1% other products

Source: U.S. Energy Information Administration

1. What are the top two products made from oil in the United States?

_____

_____

2. Which products on this list are you surprised come from oil?

_____

_____

3. What ideas would you suggest to reduce the world's use of oil?

_____

_____

Name: _____ **Date:** _____

Directions: In the United States, 65 percent of electricity is made from fossil fuels. Many cars and home heaters run on fossil fuels, too. How do you and your family use fossil fuels? Write or draw items and activities that use energy from fossil fuels.

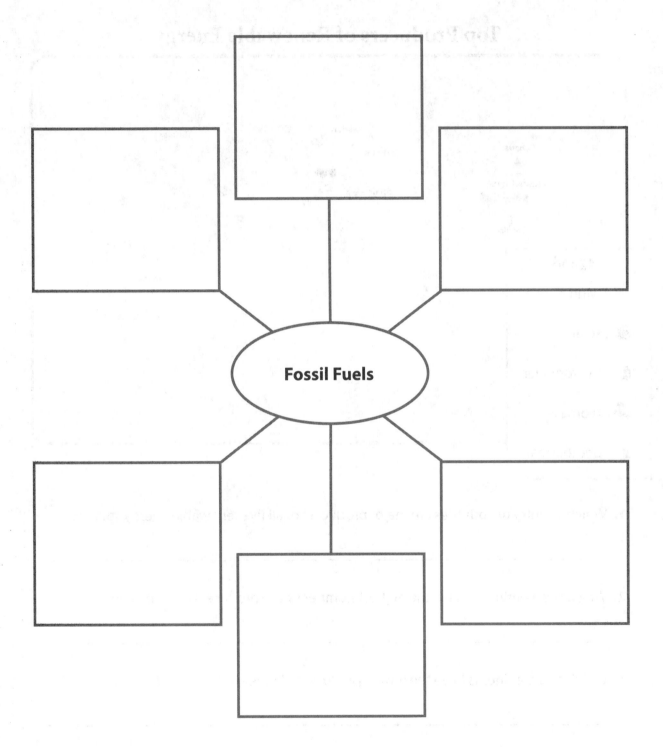

Reading Maps

Name: _____ **Date:** _____

Directions: This map shows the top producers of renewable energy. Those energy sources include wind, solar, hydropower (water), biomass (wood and waste), and geothermal (heat from Earth's crust). Study the map, and answer the questions.

Top Producers of Renewable Energy

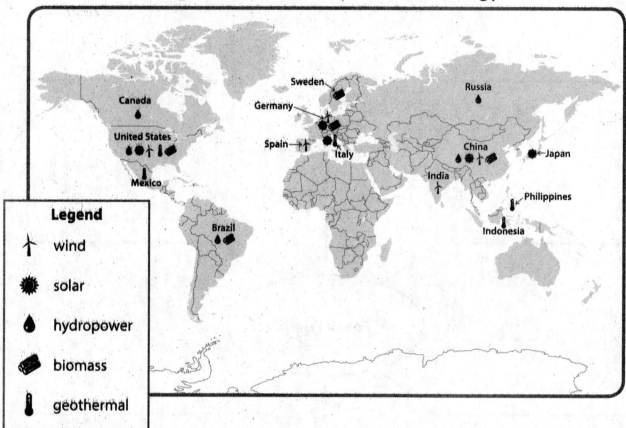

Legend

↑ wind

☀ solar

💧 hydropower

biomass

geothermal

1. Which country or countries are major producers of all five renewable energy sources?

   _____

2. Which two continents have the highest number of top producers of renewable energy?

   _____

3. Which two continents have the fewest producers of renewable energy?

   _____

Name: _____ Date: _____

Directions: Use colors or shading to represent the information in the chart. Be sure to include a legend.

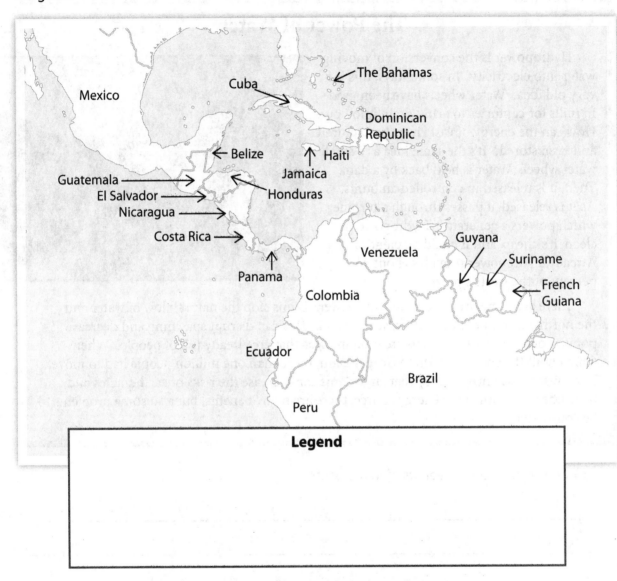

Legend

Countries with the Highest Percentage of Renewable Energy in 2012	
Country	Percentage
Ecuador	66%
Belize	63%
Panama	61%
Guatemala	60%
El Salvador	54%
Venezuela	53%
Nicaragua	52%

Name: _____ **Date:** _____

Directions: Read the text, and study the photo. Then, answer the questions.

The Power of Water

Hydropower is the conversion of moving water into electricity. In some ways, this is a very old idea. Water wheels have been used in mills for centuries to grind grain into flour. How can the energy of vast rivers be captured and even stored? It's the same idea as a small water wheel. Water is held back by a dam. Then, it is released in controlled amounts. As it is released, it passes through a turbine, which powers a generator. Hydropower is clean, it's cheap, and it could be plentiful. Already, hydropower provides about 20 percent of the world's power.

There are some drawbacks to hydropower. Dams stop the natural flow of water and the normal movement of wildlife, such as fish. This can disrupt spawning and decrease populations. Sometimes, dams are built in areas that are already full of people. When China built the enormous Three Gorges Dam, more than one million people had to move. Finally, some scientists worry that large dams can increase the risks of earthquakes and landslides. As with every energy source, there are many benefits, but also some problems to be considered.

1. How has hydropower been used in the past?

_____

_____

2. How is the energy from a river captured?

_____

_____

3. What are some problems with hydropower?

_____

_____

Name: _____ **Date:** _____

Directions: This table shows what percentage of the world's electric and plug-in hybrid cars were owned in different countries in 2015. Study the table, and answer the questions.

Country	Share
China	1.0%
France	1.2%
Netherlands	9.7%
Norway	23.3%
United States	0.7%
Italy	0.1%

1. How does the United States compare to other countries in terms of adopting electric and plug-in hybrid vehicles?

   _____

   _____

2. Why do you think the share of electric and plug-in hybrids is not higher for the United States?

   _____

   _____

   _____

3. In what ways can this table be misleading?

   _____

   _____

   _____

4. Why might someone want to own an electric or plug-in hybrid vehicle as opposed to a traditional vehicle?

   _____

   _____

   _____

Think About It

Geography and Me

Name: _____ Date: _____

Directions: Many automakers are trying to use cleaner energy in cars. Hybrid and electric cars are becoming more popular. Design an advertisement for electric cars. Include pictures, convincing information, and a catchy slogan.

Name: _____ **Date:** _____

Directions: This map shows the 10 countries that hosted the highest numbers of refugees in 2015. Study the map, and answer the questions.

Top 10 Host Countries for Refugees

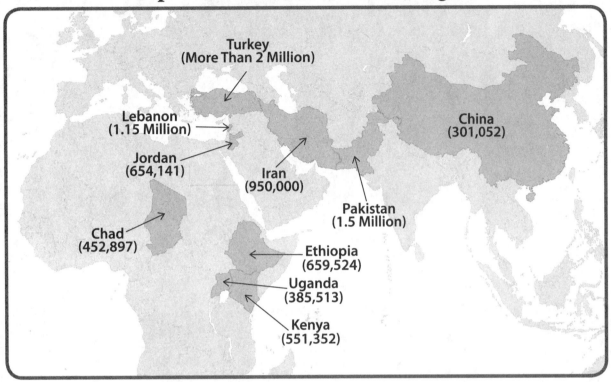

1. Which regions of the world host the most refugees?

   _____

   _____

2. Approximately how many refugees arrived in the top ten countries in 2015?

   _____

3. Approximately how many refugees arrived in the African countries listed above in 2015?

   _____

4. Why do you think refugees came to these countries?

   _____

   _____

Creating Maps

Name: _____ Date: _____

Directions: Color or shade the map to show the countries that were the sources of the most refugees in 2015. Use a map to check your answers.

Top 10 Sources of Refugees

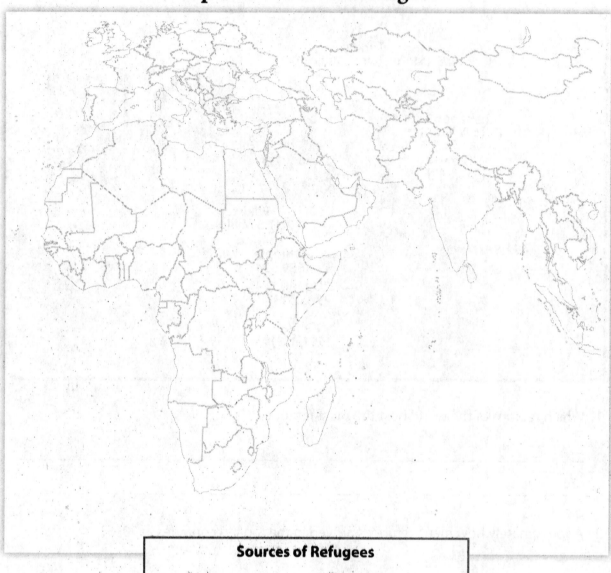

Sources of Refugees	
Syria	Afghanistan
Somalia	Sudan
South Sudan	Democratic Republic of the Congo
Myanmar	Central African Republic
Iraq	Eritrea

Name: _____ **Date:** _____

Directions: Read the text, and study the photo. Then, answer the questions.

Immigrants and Refugees

People move for all kinds of reasons. Sometimes, people move to find new jobs, better schools, or to be closer to family. These are immigrants. An immigrant is someone who chooses to move to another country. When people argue over immigration limits nowadays, they are often referring to these people.

Refugees, on the other hand, move to escape danger of some kind. War, unfair political treatment, or environmental disasters may cause this danger. Often, refugees flee to neighboring countries. Once there, they ask for asylum. *Asylum* means protection. Many countries want to help refugees in crisis. Since refugees usually arrive from nearby countries, people often feel obliged to help.

When there are many refugees, the host country, or groups within it, often set up refugee camps. These camps provide temporary housing, food, and medical support for refugees. Refugee camps are not luxury resorts. Sometimes, they are dangerous, too. Usually, the goal is to return home once the danger has passed. Sometimes, refugees end up staying for months or even years. Imagine being stuck in limbo, not knowing when you can return home. No one would choose that life.

1. What is the difference between an immigrant and a refugee?

_____

_____

2. What are some reasons why refugees might flee?

_____

_____

3. What are some of the difficulties of refugee life?

_____

_____

Think About It

Name: _____ **Date:** _____

Directions: The table on the left shows how many immigrants are living in the United States as of 2013. The table on the right shows how many people left the United States in 2013. Study the tables, and answer the questions.

Immigrants Living in the United States (2013)		People Leaving the United States (2013)	
Origin	Total	Destination	Total
Mexico	12,950,828	Mexico	848,576
China	2,246,840	Canada	316,649
India	2,060,771	United Kingdom	222,201
Philippines	1,998,932	Puerto Rico	188,954
Puerto Rico	1,685,015	Germany	111,375
Total	20,942,386	Total	1,687,755

1. Which countries or areas are listed in both sides of the table?

   _____

   _____

2. Why do you think Mexico and Canada were the top two destinations for people leaving the United States?

   _____

   _____

3. What are some of the challenges you think immigrants in the United States face, and why?

   _____

   _____

   _____

   _____

4. How are these challenges similar to and different from those facing refugees?

   _____

   _____

   _____

   _____

28627—180 Days of Geography

© Shell Education

Name: _____ **Date:** _____

Directions: Design a map of your community that might be useful to a newly arrived refugee family. Think about what they would need if they arrived with few possessions. Be sure to include a legend.

Geography and Me

Reading Maps

Name: _____ Date: _____

Directions: This map shows the world's megacities. A megacity has a population of more than 10 million people. The larger the circle is on the map, the larger the population is of the megacity. Study the map, and answer the questions.

Megacities

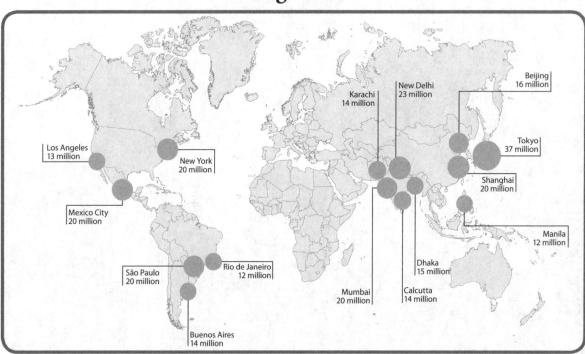

1. Which two megacities have the highest populations? What are their populations?

   _____

   _____

2. Which continent has the highest number of megacities? How many are on that continent?

   _____

3. Which region has the highest number of megacities? List each megacity in that region.

   _____

   _____

4. Describe what you may see in a megacity.

   _____

   _____

28627—*180 Days of Geography*   © *Shell Education*

Name: _____ Date: _____

Directions: Follow the steps to complete the map.

Megacities

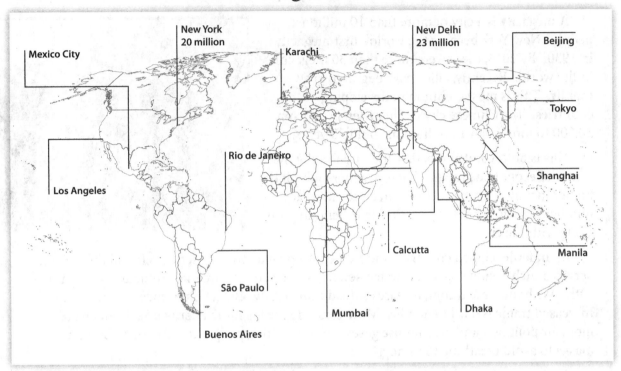

1. Color or shade the countries that will have projected megacities of 25 million or more people in 2025. The cities include: Mexico City, Mumbai, New Delhi, Tokyo, and Shanghai.

2. Add any more large cities that you know on the map.

3 Label any countries that you know on the map.

4. Place a small circle on the map to show where you live.

5. Use a map to check your answers.

Read About It

Name: _____   **Date:** _____

Directions: Read the text, and study the photo. Then, answer the questions.

Megacities and Mega-Problems

A megacity is a city of more than 10 million people. New York became the world's first megacity in 1930. By 2016, there were more than 30 megacities in the world. Recently, cities have begun to grow rapidly. This is most common in developing countries. The Chinese city of Shenzhen grew from 30,000 to more than 10 million in about 30 years.

This is both good and bad. Cities are usually the center of economic growth. There are more jobs and higher incomes in cities. However, large cities face a number of serious challenges. Problems tend to grow along with the cities.

In underdeveloped countries, megacities struggle to provide services for all the people. Services such as housing, clean water, sewers, and electricity are most difficult to keep up with. Without these things, diseases spread more easily. Bigger cities mean more cars. Increased traffic is hard to handle. More cars also means more energy use. Both create more air pollution and greenhouse gases. In some cities, people cover their faces with masks to avoid breathing the smog.

As the world population continues to grow, we should rethink megacities. One possible answer is to support the growth of smaller cities and towns. Then, people would have other places to live and work. Megacities might even shrink. That could be a better long-term solution for everyone.

1. Where are megacities growing most rapidly?

   _____

   _____

2. What are some benefits of the growth of cities, or urbanization?

   _____

   _____

3. What is the biggest problem facing megacities, and why?

   _____

   _____

Name: _____ **Date:** _____

Directions: This is a photo of solar panels for a power plant outside Shanghai, China. Study the photo, and answer the questions.

1. What problem common to all megacities are these solar panels addressing?

_____

_____

_____

2. What other ways could this problem be addressed?

_____

_____

_____

3. What is the best way for a megacity to control its population growth? Why?

_____

_____

_____

Name: _____ Date: _____

Geography and Me

Directions: What do you predict the world will look like in 50 years? Think about the rainforests, coral reefs, glaciers, and oceans. Consider population density and urbanization. Finally, consider how people and animals might adapt to any changes in the planet. Draw the world of the future. Then, summarize your drawing.

_____

_____

_____

_____

_____

28627—180 Days of Geography   © Shell Education

ANSWER KEY

There are many open-ended pages and writing prompts in this book. For those activities, the answers will vary. Answers are only given in this answer key if they are specific.

Week 1 Day 2 (page 16)

1. grassland
2. two parking lots
3. The car symbol on the bottom-right side of the map should be circled.
4. Two different routes should be drawn.

Week 1 Day 3 (page 17)

1. about 400 miles
2. Answers will vary but should include at least five different countries.

Week 1 Day 4 (page 18)

1. They increase because they are moving farther away from the prime meridian
2. about 1,000 miles
3. about 1,000 miles

Week 2 Day 3 (page 22)

1. Shading shows the different annual rainfall in China.
2. the southeastern part of the country and the two islands
3. Farmers need water for crops, so they would use the map to find the best place to farm.

Week 2 Day 5 (page 24)

1. Political because you need to know where things are. Political maps show cities, states, and borders. Political maps often show roads as well.
2. Thematic, with different colors representing each category of wealth.
3. Physical because it shows elevation and bodies of water.

Week 3 Day 1 (page 25)

1. borders between countries and the names of countries
2. Outline Botswana, Burkina Faso, Burundi, Central African Republic, Chad, Ethiopia, Lesotho, Malawi, Mali, Niger, Rwanda, South, Sudan, Swaziland, Uganda, Zambia, and Zimbabwe
3. A line should be drawn between the triangle and lake symbols.
4. approximately 1,200 miles

Week 3 Day 3 (page 27)

1. They make it seem that Africa is all the same, as if it were one country. They also make it seem like Africa is underdeveloped.
2. There are 54 countries and hundreds of different languages and cultural groups.
3. Multiple languages make education and governing more difficult. Conflicts and wars may develop between ethnic groups.

Week 3 Day 4 (page 28)

1. Urban population has increased.
2. Djibouti, Libya, and Gabon

Week 4 Day 1 (page 30)

1. The Nile flows north and empties into the Mediterranean Sea.
3. 800–900 miles.

Week 4 Day 3 (page 32)

1. There was little rainfall or other water. Floods in the Nile brought water and good soil for farming.
2. They built nilometers to track the amount of flooding each year.
3. It allowed Egypt to control the flow of the Nile. It allowed them to capture energy from the water. But it decreased the fertility of the land around the Nile.

Week 4 Day 4 (page 33)

1. 4.73 trillion cubic feet less.
2. The Aswan High Dam because it has so much more water. The Aswan High Dam does generate more than twice as much electricity as the Hoover Dam.
3. The Nile River is so large.

Week 5 Day 1 (page 35)

1. Egypt, Libya, and Algeria
2. Mali, Niger, Chad, Sudan, and Algeria
3. Laayoune, Kartoum, and Cairo

Week 5 Day 2 (page 36)

1. Cairo to Timbuktu: about 3,000 miles.
2. Tripoli to Timbuktu: about 1,800 miles
3. Tunis to Timbuktu: about 2,000 miles
4. Ceuta to Timbuktu: about 1,600 miles

ANSWER KEY (cont.)

Week 5 Day 3 (page 37)

1. It had water and was full of plants and animals.
2. The landscape had much more plant life and was greener back then than it is today.
3. They show animals that could not live in a desert climate, proving that it was not always a desert.

Week 5 Day 4 (page 38)

1. It's hard to build roads, and you must use slower methods, like camels.
2. People have to wear clothes, build houses, and uses resources that use less water and can stand up to the heat.

Week 6 Day 1 (page 40)

1. Democratic Republic of the Congo, Central African Republic, Cameroon, Republic of the Congo, and Angola
2. 1,200 miles
3. 1,300 miles

Week 6 Day 3 (page 42)

1. The Congo River Basin is dense, thick rainforest that is very large.
2. by water, there are too many waterfalls; by land, the rainforest is too thick.
3. Trade and communication could be improved strengthening the economy and government.

Week 6 Day 4 (page 43)

1. As the temperature goes up, so does rainfall, and as the temperature goes down, so does rainfall.
2. It is warm and rainy from January to May and from September to December. It is dry and cooler from June to August.
3. Answer should be supported with details from the chart.

Week 7 Day 1 (page 45)

1. Any of these: Sofala, Kilwa, Zanzibar, Mombasa, Lamu, and Mogadishu
2. Answers should include four of the following: Indian Ocean, Red Sea, Arabian Sea, Bay of Bengal, South China Sea, and Philippine Sea.
3. Three continents: Africa, Asia, and Europe.

Week 7 Day 3 (page 47)

1. the east coast of Africa, the Arabian Peninsula, South Asia, and Australia
2. It was easy to navigate because you could stay close to the coast. In addition, each region had different resources that were in demand.
3. They adopted some elements of Arab and Persian architecture, religion, and language.

Week 7 Day 4 (page 48)

1. Arabic words are about travel (travel, of the coast, friend) and products that were traded (cotton). Persian words are about products that were traded (tea and pickles).
2. Students may sort the words based on their origins or the meanings of the words.
3. Traders, sailors, and business people would use these words.
4. Some were items that were new to the area, so they had no name in Swahili before then. Others related to activities that were new with trade, so they probably had no name before either.

Week 8 Day 1 (page 50)

1. Russia, China, India.
2. Turkey is farthest to the west. Russia extends farthest to the east.
3. Russia extends up into the Arctic Circle. Indonesia and East Timor extend below the equator.
4. China

Week 8 Day 3 (page 52)

1. a combination of geography, culture, and history
2. They are all on the Arabian Peninsula, they trace their ancestors back to early civilizations in that area, and many share religion and language.
3. Yes Israel is an example because it is part of the Middle East but does not share the same culture, language or religion as other nations in the region.

Week 8 Day 4 (page 53)

1. Most regions are increasing in population.
2. Overall population in 2015 was 4,536,753. That was an increase of 675,882 from 2000
3. In 2000, it was Eastern Asia.
4. In 2015, it was South Asia.
5. Yes, population is increasing everywhere because of the birth rate.

ANSWER KEY (cont.)

Week 9 Day 1 (page 55)

1. Mediterranean Sea, Red Sea, Persian Gulf, Arabian Peninsula, Nile River, Tigris River, Zagros Mountains, Syrian Desert, and Euphrates River
2. Israel, Lebanon, Syria, Turkey, Iraq, and Iran
3. the Tigris and Euphrates Rivers because people need water to live and farm

Week 9 Day 3 (page 57)

1. for drinking water and farming
2. The flooding of the Tigris and Euphrates rivers was not predictable.
3. First, they built basins, and then they built canals.

Week 9 Day 4 (page 58)

1. East and Southeast Asia
2. Egypt, the Middle East, India/Pakistan, and China.
3. Egypt and Mesopotamia
4. There was likely trade between Egypt and Mesopotamia fairly early, based on their proximity and easy access by land or water. Exchange of ideas and good.

Week 10 Day 1 (page 60)

1. summer; The monsoons are blowing towards India.
2. south
3. Northern China, India, and Nepal
4. Indonesia and Southeast Asian countries

Week 10 Day 3 (page 62)

1. They are systems of seasonal windstorms and are associated with heavy rains.
2. Monsoons determined when and where traders go. They also affected farming due to rainfall and temperature.
3. Flooding can cause landslides, the spread of diseases, property damage, and death.

Week 10 Day 4 (page 63)

1. When rainfall increases, malaria has an almost identical increase.
2. They tend to match fairly well during the rainy season. But the numbers rise before the rains come, which means there must be other factors at play.
3. They need to find a way to address the accumulation of water because it contributes to significant health problems. For malaria, they should address the mosquito population or give out preventative medicine.

Week 11 Day 1 (page 65)

1. Kathmandu
2. approximately 100 miles, 160 kilometers
3. They are located along the coast.
4. Answers may include that they have good harbors or that they are close to the ocean, a major resource.

Week 11 Day 3 (page 67)

1. It's hard to keep producing enough resources to match rapid growth.
2. They encouraged people to have smaller families. They used an advertising campaign and even cash payments.
3. It denies freedom to its people but can help save resources.

Week 11 Day 4 (page 68)

1. highest: Singapore; lowest: Sri Lanka
2. Urbanization is happening faster in Southeastern Asia than in Southern Asia.

Week 12 Day 1 (page 70)

1. India
2. Answers may include China and Japan.
3. There is a lot of religious diversity in Asia.

ANSWER KEY *(cont.)*

Week 12 Day 3 (page 72)

1. They come from the same area, and holy city, Jerusalem. They are monotheistic.
2. Judaism spreads when its followers migrate and new generations are born. Often in history, Judaism spread to a new area as its followers moved to escape persecution.
3. They both spread early through conquest and missionary activity.

Week 12 Day 4 (page 73)

1. Hinduism in 2010; Islam in 2050
2. Birth rates might be different among different countries or among different religious groups.
3. All of these major world religions originated in Asia, so it makes sense that they would all be represented there. From there, they have spread more or less, depending on many circumstances.

Week 13 Day 1 (page 75)

1. Northern Mariana Islands, Guam, Wake Island, Midway Islands, Kingman Reef, Howland Island, Johnston Atoll, Palmyra Atoll, Jarvis Island, Baker Island, Hawai'i, and American Samoa
2. about 1,500 miles
3. Fiji: 20°S,180°; Guam: 15°N,145°E; Easter Island: 25°S,110°W

Week 13 Day 3 (page 77)

1. They came from Southeast Asia around 1500 BC.
2. They couldn't understand how the Polynesians could have traveled so far at sea with only canoes and no compasses.
3. They built replica canoes and tried sailing according to advice given by Polynesians centuries before. It worked.

Week 13 Day 4 (page 78)

1. Polynesians were great sailors and navigators of the ocean.
2. storms, distance, and the small size of the canoes
3. There is much the world can learn from every culture, so they should not be ignored.

Week 14 Day 1 (page 80)

1. Zealandia: 2,100 miles long, 1,500 miles wide
   Australia: 1,500 miles long, 1,800 miles wide
2. the southern end of the continent
3. 300 miles

Week 14 Day 3 (page 82)

1. Not in the traditional sense. It covers a large area of land, but most of it is far underwater.
2. Oil companies began drilling deep in the water near New Zealand, where they discovered continental rocks. Scientists later found much more.
3. If New Zealand can prove that Zealandia is part of the same continent, they will get much more mining and fishing rights. That means money.

Week 14 Day 4 (page 83)

1. The size or location of a piece of land as the key characteristic of a continent.
2. Predictions will vary. Students might expect some pushback, since most of the land is so far underwater.
3. There may be valuable resources in Zealandia.

Week 15 Day 1 (page 85)

1. New South Wales, Northern Territory, Queensland, South Australia, Victoria, Tasmania and Western Australia
2. New South Wales and Queensland
3. Indian Ocean, Timor Sea, Arafura Sea, Gulf of Carpentaria, Coral Sea, Tasman Sea, and Great Australian Bight

Week 15 Day 3 (page 87)

1. It is more than 70% of Australia, and the land includes desert, forest, mountains, tropical grasslands, scrub, and wetlands.
2. It is isolated with few people.
3. They managed fires, and they dug waterholes.

Week 15 Day 4 (page 88)

1. convicts and soldiers
2. organization, control, and lack of resources
3. It is very far away and isolated.

Week 16 Day 1 (page 90)

1. forest and woodland
2. mostly in the center or northeast part of the country
3. shrubland
4. forest

ANSWER KEY *(cont.)*

Week 16 Day 3 (page 92)

1. They are both marsupials.
2. Dingoes are everywhere, and they eat a lot of livestock.
3. They look like large rats. They are mean and angry.

Week 16 Day 4 (page 93)

1. Fish – 1.5%, Mammals – 1.8%, Birds – 1.1%, Plants—0.06%
2. Australia has the highest number of fish on the threatened species list because of The Great Barrier Reef.
3. Plants will get the least attention because they are not as interesting to most people as animals.
4. Answers may include fining people for disturbing these species, protecting their habitats, and funding research on the species.

Week 17 Day 1 (page 95)

1. Asia, North America, and Africa
2. Caribbean Sea, Indian Ocean, Red Sea, Pacific Ocean
3. between the Tropic of Cancer and the Tropic of Capricorn

Week 17 Day 3 (page 97)

1. Coral reefs are underwater ridges of coral, seaweed, and sea sponges.
2. Humans are overfishing, fishing with dynamite, accidentally damaging while diving or boating, and littering.
3. Coral bleaching is the whitening of coral due to rising ocean temperatures. Bleaching can lead to the death of the coral.

Week 17 Day 4 (page 98)

1. They may intentionally or accidentally touch the coral and damage it.
2. submarine, snorkeling, glass bottom boats

Week 18 Day 1 (page 100)

1. Armenia and Mesopotamia are farthest east. Lusitania is farthest west.
2. The Roman Empire covered parts of Europe, Asia, and Africa and was centered around the Mediterranean Sea.
3. Possible answers are Britannia, Italia, Africa, Germania, Syria, Armenia, Belgica, Macedonia, and Aegyptus.

Week 18 Day 3 (page 102)

1. They were located right next to each other, sharing some of their territory. They also overlapped in time.
2. They demonstrate how Ancient Romans adopted and used Greek architectural forms. The Pantheon uses Corinthian columns.
3. They conquered much of Europe, establishing settlements and taking control of the political power.

Week 18 Day 4 (page 103)

1. They are names of planets in our solar system.
2. Answers may include that its large size made it seem powerful.
3. The ancient Greeks and Romans valued family and love, war and conquest, food, and travel.

Week 19 Day 1 (page 105)

1. There are seven on this map: Campi Flegrei, Vesuvius, Stromboli, Panarea, Etna, Campi Felgrei Del Mar Di Sicilia
2. Naples and Palermo
3. All the major volcanoes in Italy are close to the plate boundaries.

Week 19 Day 3 (page 107)

1. It is an active volcano in Italy.
2. Vesuvius erupted and destroyed the nearby cities of Herculaneum, Stabiae, and Pompeii.
3. Researchers are still studying this site because of the preservation by the volcano.
4. It was covered with so much debris, so quickly. The debris served as insulation.

ANSWER KEY *(cont.)*

Week 19 Day 4 (page 108)

1. Most volcanoes have not been active in decades. Some have been dormant for centuries. There are two volcanoes that seem to be currently active: Etna and Stromboli.
2. Most of the volcanoes that have erupted in the last few centuries have been stratovolcanoes. Students might also notice that the three most recently erupted volcanoes were the three with the highest elevation.
3. Volcanologists and officials need to plan for evacuation and notification procedures to reach everyone who might be in harm's way in case of an eruption.
4. Answers may include warning people about the potential dangers or funding additional research.

Week 20 Day 1 (page 110)

1. Norway
2. It should be cold, with plenty of snow and ice.
3. The coastline is wavy or jagged, not smooth.

Week 20 Day 3 (page 112)

1. A fjord is a deep, narrow waterway where the ocean stretches inland.
2. They were formed slowly during the ice age by glaciers.
3. Visitors enjoy the beauty and hike, fish, and kayak through the fjords.

Week 20 Day 4 (page 113)

1. Answers may include that the land is steep, jagged, and surrounded by many other fjords.
2. Answers may include the beauty and the access to fishing.
3. Answers may include that the culture might remain traditional because of the isolation.
4. Erosion from ice, rain, and wind as well as pollution and building affect fjords.

Week 21 Day 1 (page 115)

1. From west to east and south.
2. 87 years (1917–1830)
3. The exception to the pattern was Italy which had railroads in 1839.

Week 21 Day 3 (page 117)

1. Britain
2. Trade and industry benefited the most.
3. People worried about safety and environmental damage.
4. Trains are still popular and are much faster.

Week 21 Day 4 (page 118)

1. France for both
2. Most countries in Western Europe increased the number of passenger-kilometers. Most countries in Eastern Europe decreased 1980–2015.

Week 22 Day 1 (page 120)

1. Northern Ireland, England, Scotland and Wales
2. The Prime Meridian runs through the eastern part of England.
3. The distance between France and England is about 20 miles at its closest.
4. The distance between France and England is about 130 miles at its furthest.

Week 22 Day 3 (page 122)

1. Some men are wearing Scottish kilts. Others are carrying Scottish flags.
2. Independent Scotland would be richer, and Scots should rule Scotland.
3. Scotland would no longer be part of the United Kingdom.

Week 22 Day 4 (page 123)

1. About 65% said Scottish only; about 8% said British only.
2. Joint Scottish and British was more common.
3. Yes, because so many people had a Scottish-only identity.

Week 23 Day 1 (page 125)

1. Answers should be close to 1,600 miles.
2. It is hot and rainy like a tropical rainforest.
3. Mexico
4. Honduras

Week 23 Day 3 (page 127)

1. It was safe and had fertile land.
2. The Spaniards drained it in a failed attempt to use it for irrigation.
3. The people of Mexico City have a shortage of water and sewage problems. Also, their city is sinking.

ANSWER KEY *(cont.)*

Week 23 Day 4 (page 128)

1. There is water surrounding the city and islands.
2. The artist seemed impressed and surprised. The drawing depicts orderly streets, a large population, and sacrificial practices.

Week 24 Day 1 (page 130)

1. 60° N, British Columbia, Alberta, Saskatchewan, Manitoba
2. Most of the population is located in Ontario and Quebec because most of the major cities are there near the Great Lakes.
3. 49°N

Week 24 Day 3 (page 132)

1. It is the eighth largest lake in the world, and it was named a UNESCO Biosphere Reserve.
2. He said that Great Bear Lake would be one of the last refuges left on Earth near the end of the world.
3. Answers may include lakes or rivers because they provide freshwater that is easy to access.

Week 24 Day 4 (page 133)

1. It has the third highest amount.
2. Answers may include that there is ample freshwater available for the population.
3. Egypt; The freshwater resources are being used up too quickly to be replenished.

Week 25 Day 1 (page 135)

1. They need warmer weather, so they migrate to the warm regions during the colder months.
2. Answers may include that they can only live in warm areas because they go to tropical warm, locations during winter.
3. They avoid mountains because it is colder in the mountains.

Week 25 Day 3 (page 137)

1. habitat destruction and the decline in milkweed.
2. It's their favorite food and where they like to lay their eggs.
3. People can encourage their towns to plant milkweed along the roadside, restrict the use of herbicides that kill milkweed, and plant milkweed in our own backyards.

Week 25 Day 4 (page 138)

1. It tends to go up and down, but the trend was down until 2016.
2. The sharpest decrease was from 1997 to 1998 Reasons may include a change in the weather or habitat.
3. Hopefully, the numbers will continue to increase. It depends largely on how much milkweed is available to them.

Week 26 Day 1 (page 140)

1. They are all islands or portions of an island.
2. They are territories of other countries, including the U.K., France, the Netherlands, and the U.S.
3. Cuba is the largest, followed by Haiti, the Dominican Republic, Jamaica, and the Bahamas. (Puerto Rico is a US territory)..

Week 26 Day 3 (page 142)

1. injuries, loss of life, property damage, flooding, and wind damage
2. They are islands that often lie in the path of hurricanes.
3. Get an emergency kit and generator, cover your windows, get an emergency radio, and follow instructions. Evacuate when necessary.

Week 26 Day 4 (page 143)

1. The houses and cars have been smashed by debris.
2. They might become homeless, lose their car, have no way to get around, be stranded and in need of medical attention, etc.
3. Answers may include fallen trees or debris.
4. Answers may include that they can uproot trees and destroy property.

Week 27 Day 1 (page 145)

1. Baja California and Yucatan Peninsula
2. Baja California is surrounded by the Pacific Ocean and the Sea of Cortez. The Yucatan Peninsula is surrounded by the Caribbean Sea and the Gulf of Mexico.
3. Lake Nicaragua in Nicaragua
4. along the western, eastern, and southern coasts of Mexico

ANSWER KEY *(cont.)*

Week 27 Day 3 (page 147)

1. coffee, fruit, and sugar
2. They were criticized for environmental damage, worker exploitation, and interference in the government.

Week 27 Day 4 (page 148)

1. Except for Panama, they all increased steadily. Some increases were more than ten times the value in 1985.
2. Honduras increased the most: 4,759,684,807 – 374,400,000 = $4,384,284,807.
3. Reasons could be changes within these countries, like more/better production, new items to sell, and higher prices, and the end of war, or changes in the U.S., like more demand for products,or lack of other choices.
4. Answers should indicate that the amount of exports will continue to increase.

Week 28 Day 1 (page 150)

1. Brazil
2. Andes Mountains
3. 3,300 miles

Week 28 Day 3 (page 152)

1. It's a series of roads connecting North and South America, almost all the way from the bottom to the top.
2. There is opposition in Darien because road construction would go through wild conservation lands.
3. It makes tourist travel and trade easier between the countries of North and South America. It also connects some areas that might be quite isolated otherwise.

Week 28 Day 4 (page 153)

1. It is half or less the size of Asia or Africa. It is only larger than two inhabited continents: Europe and Australia.
2. It means that on average, countries will be bigger in South America than elsewhere, and there will probably be some very large countries.
3. Answers may include that since South America is a large continent, it is very diverse.

Week 29 Day 1 (page 155)

1. The main river runs through Brazil and Peru.
2. Brazil, French Guiana (a French territory), Suriname, Guyana, Venezuela, Colombia, Peru, Ecuador, and Bolivia
3. 4,200 miles

Week 29 Day 3 (page 157)

1. It's the largest rainforest in the world. It has millions of species, some of which only live in the Amazon. It also holds a lot of carbon.
2. cattle ranching and large-scale farming
3. It could disrupt the ecosystem in Brazil and effect the world's climate.

Week 29 Day 4 (page 158)

1. About 18.1% was lost.
2. Conservation efforts may have been working to slow deforestation.
3. a gradual increase in the annual rate of forest loss
4. Answers may include limiting the amount of rainforest that can be cleared or replanting trees.
5. Answers may include spreading the word or writing to leaders.

Week 30 Day 1 (page 160)

1. the Andes Mountains
2. northeastern
3. Maranon, Ucuyali,Huallaga, Amazon, Piedras, Napo.

Week 30 Day 3 (page 162)

1. bad soil, steep land, and little water
2. They built canals, cisterns, and terraces. They also improved the soil to retain water.
3. There is a highway but not all paved, and there are trains.

Week 30 Day 4 (page 163)

1. Chile, Peru, and Ecuador.
2. Bolivia, Colombia, and Peru.
3. Most likely Bolivia, Colombia, and Peru have the most cities in the Andes. These countries all have a lot of their population in the mountains, and cities have the highest populations.

Week 31 Day 1 (page 165)

1. capital goods like electronics and machinery
2. oil; the Middle East, Russia, and Africa
3. Peru and Chile

ANSWER KEY *(cont.)*

Week 31 Day 2 (page 166)

3. Germany: 8,500 miles
   Italy: 8,000 miles
   United States: 5,000 miles
   China: 10,000 miles
   South Korea: 11,000 miles
4. Answers may include that Chile has a large international trade network.

Week 31 Day 3 (page 167)

1. It makes up about half of Chile's exports.
2. Chile produces more than any other country. Also, Chile is home to 6 of the top 10 largest copper mines.
3. deforestation, use of a lot of water, pollution

Week 31 Day 4 (page 168)

1. Not always stable.
2. This makes things unpredictable for Chile's economy because copper is such a large percentage of the economy.

Week 32 Day 1 (page 170)

1. Pacific Ocean
2. Isabela
3. Rabida
4. They are Spanish names and were influenced by Spanish colonization.

Week 32 Day 3 (page 172)

1. They are in the middle of the ocean and were untouched for a long time.
2. They have physical features that make sense for the land they live on.
3. Lots of tourists visit, there is construction of new buildings and amenities, and non-native animals try to destroy the natural environment.

Week 32 Day 4 (page 173)

1. They generate trash, which can damage the ecosystem.
2. They have limited space to store it, few recycling operations, and burning is likely to happen close to humans or wildlife. The Galápagos have an extra challenge because they need to preserve the habitat for their unique wildlife.
3. The animals may try to eat some of the garbage and get sick. Also, the garbage may attract rodents and other non-native pests that threaten the wildlife.

Week 33 Day 1 (page 175)

1. United States, Saudi Arabia, and Russia
2. Asia (including the Middle East)
3. Answers may include that oil is a large part of the world economy because many countries produce oil.

Week 33 Day 3 (page 177)

1. Fossil fuels are coal, oil, and natural gas. They were once plants and animals who became fossilized and buried.
2. They are plentiful and cheap. We already have systems to extract them and distribute them.
3. They pollute the air and release lots of carbon dioxide, which contributes to global warming.

Week 33 Day 4 (page 178)

1. finished motor gasoline and heating oil/fuel oil

Week 34 Day 1 (page 180)

1. The United States is the only one.
2. Asia and Europe
3. Africa has none. Australia/Oceania and South America each have one.

Week 34 Day 3 (page 182)

1. Water wheels were used to power mills.
2. Dams hold back the water and release it in a controlled way. The released water spins a turbine, which powers a generator.
3. It disrupts the flow of rivers, which disrupts the habitat and behaviors of some animals. It also sometimes displaces large numbers of people. Finally, it may increase the risk of earthquakes and landslides.

Week 34 Day 4 (page 183)

1. The U.S. is behind almost all the countries, especially Norway.
2. People are slow to change, these cars are not the cheapest, and many people need to travel longer distances.
3. The smaller size of these countries might be a factor. Governments might be providing more incentives to buy cars like these. Prices might be closer to similar-sized vehicles in these countries.
4. Answers may include that they are cleaner or cost less to fuel.

ANSWER KEY *(cont.)*

Week 35 Day 1 (page 185)

1. the Middle East, Central Asia, and Africa
2. about 8.6 million refugees
3. about 2.05 million refugees
4. Answers may include that these countries are near the ones they are fleeing from or that these countries are better off than the ones they left.

Week 35 Day 3 (page 187)

1. Immigrants move willingly. Refugees are fleeing danger.
2. natural disasters, war, unfair political treatment
3. You have to flee your home, live in a camp that might be dangerous, and wait for an undetermined amount of time before returning home.

Week 35 Day 4 (page 188)

1. Mexico and Puerto Rico
2. They are neighboring countries.
3. economic challenges, language barriers, cultural adjustments, loneliness
4. Answers may include that they both leave home and must adapt to a new country, however, refugees may not know if or when they can return home.

Week 36 Day 1 (page 190)

1. Tokyo (37 million) and New Delhi (23 million)
2. Asia with nine cities
3. India; Karachi, New Delhi, Calcutta, and Mumbai
4. Answers may include tall buildings, crowded streets, and busy roads.

Week 36 Day 3 (page 192)

1. in developing countries
2. Cities provide jobs, stimulate economic growth, and provide higher incomes.
3. ack of decent housing, clean water, sewers, and electricity; spread of diseases; more traffic; air pollution and smog

Week 36 Day 4 (page 193)

1. It is addressing the problem of pollution and increased energy needs.
2. Cities could build more windfarms, hydropower plants, or other renewable energy sources. Cities could also encourage energy conservation and efficiency.

POLITICAL MAP OF THE UNITED STATES

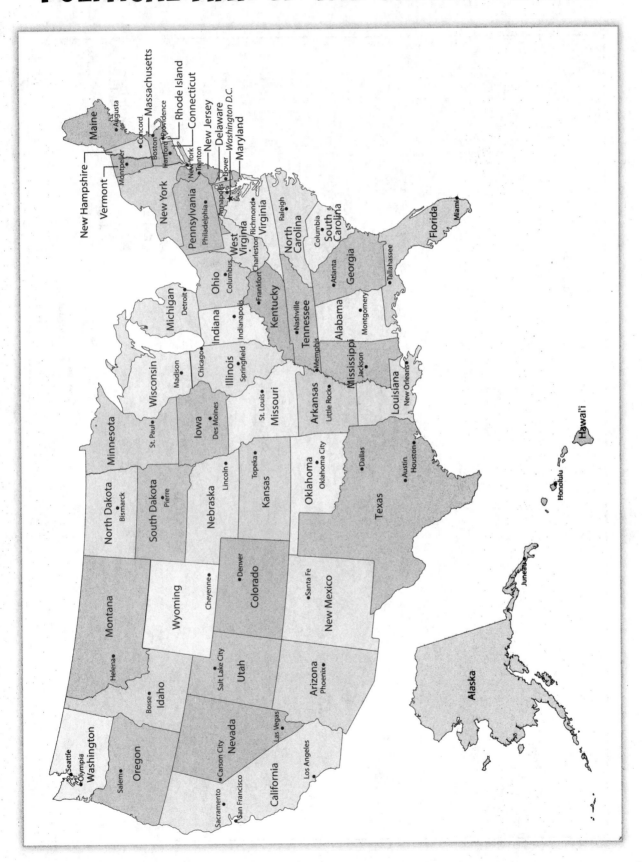

PHYSICAL MAP OF THE UNITED STATES

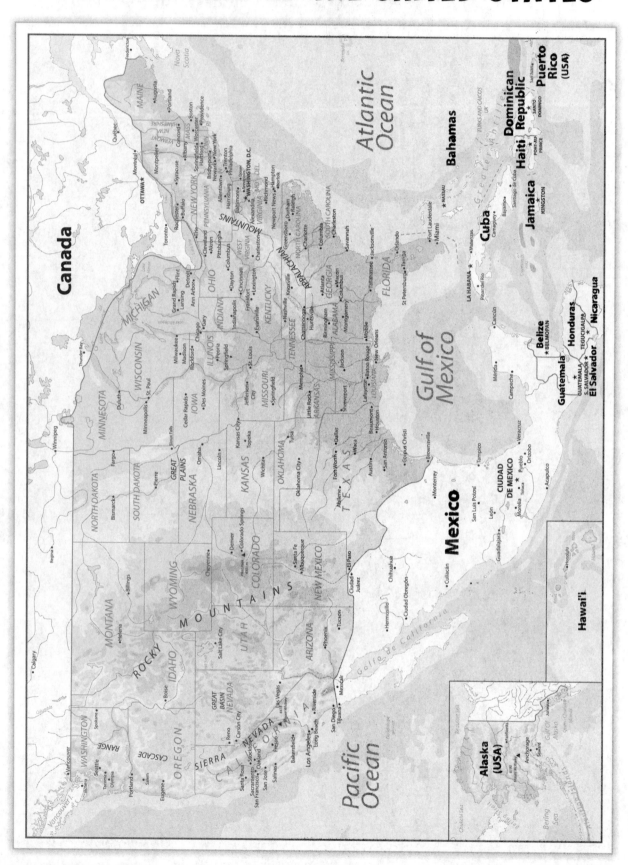

28627—180 Days of Geography

© Shell Education

WORLD MAP

WESTERN HEMISPHERE

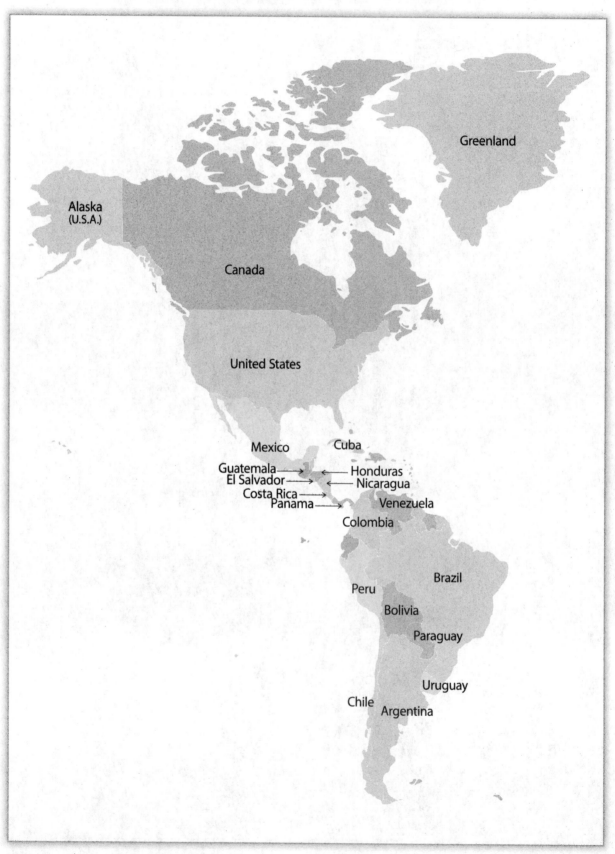

EASTERN HEMISPHERE

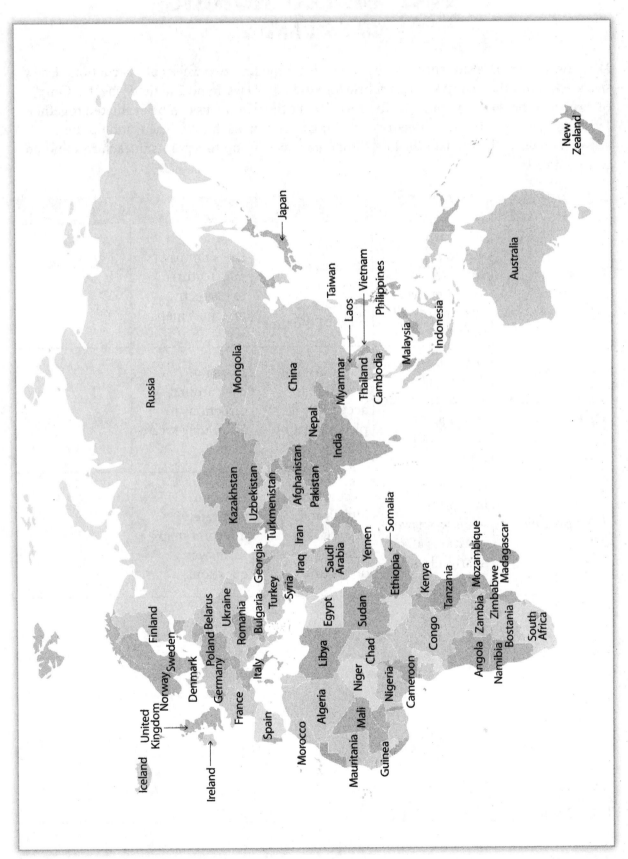

Name: _____ **Date:** _____

MAP SKILLS RUBRIC
DAYS 1 AND 2

Directions: Evaluate students' activity sheets from the first two weeks of instruction. Every five weeks after that, complete this rubric for students' Days 1 and 2 activity sheets. Only one rubric is needed per student. Their work over the five weeks can be evaluated together. Evaluate their work in each category by writing a score in each row. Then, add up their scores, and write the total on the line. Students may earn up to 5 points in each row and up to 15 points total.

Skill	5	3	1	Score
Using Map Features	Uses map features to correctly interpret maps all or nearly all the time.	Uses map features to correctly interpret maps most of the time.	Does not use map features to correctly interpret maps.	
Using Cardinal Directions	Uses cardinal direction to accurately locate places all or nearly all the time.	Uses cardinal direction to accurately locate places most of the time.	Does not use cardinal directions to accurately locate places.	
Interpreting Maps	Accurately interprets maps to answer questions all or nearly all the time.	Accurately interprets maps to answer questions most of the time.	Does not accurately interpret maps to answer questions.	

Total Points: _____

Name: _____ Date: _____

APPLYING INFORMATION AND DATA RUBRIC
DAYS 3 AND 4

Directions: Complete this rubric every five weeks to evaluate students' Day 3 and Day 4 activity sheets. Only one rubric is needed per student. Their work over the five weeks can be evaluated together. Evaluate their work in each category by writing a score in each row. Then, add up their scores, and write the total on the line. Students may earn up to 5 points in each row and up to 15 points total. **Note:** Weeks 1 and 2 are map skills only and will not be evaluated here.

Skill	5	3	1	Score
Interpreting Text	Correctly interprets texts to answer questions all or nearly all the time.	Correctly interprets texts to answer questions most of the time.	Does not correctly interpret texts to answer questions.	
Interpreting Data	Correctly interprets data to answer questions all or nearly all the time.	Correctly interprets data to answer questions most of the time.	Does not correctly interpret data to answer questions.	
Applying Information	Applies new information and data to known information about locations or regions all or nearly all the time.	Applies new information and data to known information about locations or regions most of the time.	Does not apply new information and data to known information about locations or regions.	

Total Points: _____

MAKING CONNECTIONS RUBRIC
DAY 5

Directions: Complete this rubric every five weeks to evaluate students' Day 5 activity sheets. Only one rubric is needed per student. Their work over the five weeks can be evaluated together. Evaluate their work in each category by writing a score in each row. Then, add up their scores, and write the total on the line. Students may earn up to 5 points in each row and up to 15 points total. **Note:** Weeks 1 and 2 are map skills only and will not be evaluated here.

Skill	5	3	1	Score
Comparing One's Community	Makes meaningful comparisons of one's own home or community to others all or nearly all the time.	Makes meaningful comparisons of one's own home or community to others most of the time.	Does not make meaningful comparisons of one's own home or community to others.	
Comparing One's Life	Makes meaningful comparisons of one's daily life to those in other locations or regions all or nearly all the time.	Makes meaningful comparisons of one's daily life to those in other locations or regions most of the time.	Does not make meaningful comparisons of one's daily life to those in other locations or regions.	
Making Connections	Uses information about other locations or region to make meaningful connections about life there all or nearly all the time.	Uses information about locations or regions to make meaningful connections about life there most of the time.	Does not use information about locations or regions to make meaningful connections about life there.	

Total Points: _____

MAP SKILLS ANALYSIS

Directions: Record each student's rubric scores (page 210) in the appropriate columns. Add the totals, and record the sums in the Total Scores column. Record the average class score in the last row. You can view: (1) which students are not understanding map skills and (2) how students progress throughout the school year.

Student Name	Week 2	Week 7	Week 12	Week 17	Week 22	Week 27	Week 32	Week 36	Total Scores
Average Classroom Score									

APPLYING INFORMATION
AND DATA ANALYSIS

Directions: Record each student's rubric scores (page 211) in the appropriate columns. Add the totals, and record the sums in the Total Scores column. Record the average class score in the last row. You can view: (1) which students are not understanding how to analyze information and data and (2) how students progress throughout the school year.

Student Name	Week 7	Week 12	Week 17	Week 22	Week 27	Week 32	Week 36	Total Scores
Average Classroom Score								

MAKING CONNECTIONS ANALYSIS

Directions: Record each student's rubric scores (page 212) in the appropriate columns. Add the totals, and record the sums in the Total Scores column. Record the average class score in the last row. You can view: (1) which students are not understanding how to make connections to geography and (2) how students progress throughout the school year.

Student Name	Week 7	Week 12	Week 17	Week 22	Week 27	Week 32	Week 36	Total Scores
Average Classroom Score								

DIGITAL RESOURCES

To access the digital resources, go to this website and enter the following code: 53196220.
www.teachercreatedmaterials.com/administrators/download-files/

Rubrics

Resource	Filename
Map Skills Rubric	skillsrubric.pdf
Applying Information and Data Rubric	datarubric.pdf
Making Connections Rubric	connectrubric.pdf

Item Analysis Sheets

Resource	Filename
Map Skills Analysis	skillsanalysis.pdf
	skillsanalysis.docx
	skillsanalysis.xlsx
Applying Information and Data Analysis	dataanalysis.pdf
	dataanalysis.docx
	dataanalysis.xlsx
Making Connections Analysis	connectanalysis.pdf
	connectanalysis.docx
	connectanalysis.xlsx

Standards

Resource	Filename
Standards Charts	standards.pdf